Reflections on Hoshin Planning

GUIDANCE FOR LEADERS AND PRACTITIONERS

Reflections on Hoshin Planning

GUIDANCE FOR LEADERS AND PRACTITIONERS

Lisa Boisvert

CRC Press
Taylor & Francis Group
Boca Raton London New York

CRC Press is an imprint of the
Taylor & Francis Group, an **informa** business
A PRODUCTIVITY PRESS BOOK

CRC Press
Taylor & Francis Group
6000 Broken Sound Parkway NW, Suite 300
Boca Raton, FL 33487-2742

Printed on acid-free paper
Version Date: 20151120

International Standard Book Number-13: 978-1-4822-9978-6 (Paperback)

Visit the Taylor & Francis Web site at
http://www.taylorandfrancis.com

and the CRC Press Web site at
http://www.crcpress.com

Contents

Acknowledgments

Thank you to the engineers, nurses, scientists, process improvement experts, machinists, doctors, executives—all those who have invited me onto their plant floors, through their healthcare facilities, along their manufacturing lines, and into their boardrooms to share in the work of using Hoshin Planning and Process Improvement to shape the future of their organizations and to improve the quality of their employees' and customers' day-to-day lives. I've been fortunate to meet many of the fine people who dedicate their professional lives to making their respective areas of the working world better places, in the most forward-thinking and systematic way they can.

Thank you to Mike Cowley and Lois Gold for making time to read the manuscript and provide timely and high-quality feedback. You contributed important insights, on both technical and organizational issues related to implementing Hoshin Planning.

Thank you to Nick Governale for his exchanges with me early in the writing that helped me focus the article "Hoshin Disruption."

Special thanks to Kevin Geraghty for inviting me to work with him and his team as he led the Hoshin Planning at NV Energy's Generation division and for sharing the team's story and examples for use in this book.

Thank you to Julia O'Neill for inviting me to join her in facilitating the Hoshin work of her team at Merck, and for sharing parts of that plan for use in this book.

For listening and talking through the charms and challenges of writing a book and for always cheering me on, I thank Elisabeth Boisvert.

For drawing me away from the book when I needed a break and providing reality checks when my perspective wavered, I thank John Minkle.

For her friendship, intellectual repartee, and technical counsel, I thank Rebecca Xiong.

Thank you Melina Melvin for transforming many of the figures from my messy drawings and old pictures into clear and high-quality images.

The following clients and colleagues made time on the phone, in person, and by e-mail to share their stories and opinions as I wrote the book. Thank you. I enjoyed these conversations very much—you brought important intellectual and professional energy throughout the year or so it took to complete the book.

Mike Anderson, MD Dana Farber Community Cancer Care	Donny Askin TechTurn Consulting, Inc.	Geoffrey Crabtree Methodist Health System
Bob Darin CVS Caremark	Elizabeth Doty, PhD Leadership Momentum	John Frankovich NV Energy
John Hamilton Philips and PlanBase, Inc.	John Henderson Gerber Scientific, Inc.	Don Hopper NV Energy
Malcolm Isley Greenville Healthcare System	Elizabeth Liebow The Dana-Farber Cancer Institute	Paula Martin Matrix Management Institute
Glenn Mazur QFD Institute	Bob McNally Northeast Foods	David Moody, PhD Milliken Chemical
Steve Page NV Energy	Dariusz Rekowski NV Energy	Jim Robinson Merck, Inc.
Tim Rowe Cambridge Innovation Center	Peter Steinbrenner NV Energy	Rich Winters NV Energy
Carmin Quirion Wyman Advocates, Inc.	Rebecca Xiong SocMetrics	

Author

Lisa Boisvert has consulted with organizations in the application of Process Improvement and Hoshin Planning about such things as electrical power generation; the manufacture of vaccines, laser fabric cutters, roofing material, industrial surface conditioning brushes, and insulin testing kits; the processing of chicken and bread; banking; and the delivery of patient care and human services.

Prior to forming Business Centered Learning™, LLC in 1999, Boisvert served as general manager for the services division of GOAL/QPC. Earlier, as the company's research director, she convened and published the collective work of leaders from around the world in the application of Process Improvement and Hoshin Planning regarding manufacturing, healthcare, and financial services industries. Her work with these early Hoshin practitioners, progressive executives, and quality management subject matter experts continues to inform her work with leadership teams today.

Introduction

Intentions of the Book

With this book I'd like to join the written dialogue around how to leverage the special, sometimes subtle features of Hoshin Planning to advance the mission of an organization. I'd also like to consider ways to mitigate the challenges we have when applying the well-seasoned practices of Hoshin Planning under modern conditions. And it would please me if someone who has been pushed into or happily engaged in the practice of strategic and Hoshin Planning experienced a bit of relief from the stress of it by reading about the experiences of others and the consistent and acceptable imperfection of its application.

I appreciate the excellent and comprehensive technical books that have built a sound body of knowledge around Hoshin Planning history, steps, and tools. Two favorites of mine are Mike Cowley and Ellen Domb's *Beyond Strategic Vision: Effective Corporate Action with Hoshin Planning* (2011) and Thomas Jackson's *Hoshin Planning for the Lean Enterprise* (2006). Both books provide excellent instruction on how to move from beginning to end through the Hoshin Planning process, with examples and pragmatic technical advice.

In contrast to the how-to books, this book discusses a few specific application topics and environments about which I have some observations from my work with clients. I'd like to add to the dialogue about why good knowledge of the mechanics, techniques, and tools of Hoshin isn't always enough to get the job done well. The book looks at Hoshin Planning a little less through the controllable lens of methodology and a little more through the sometimes shifting and imperfect lens of daily life in organizations.

Organizations experience unplanned events and surprising results that can require the frequent modification of what look like perfectly good plans. For example, can we know for sure when starting Hoshin that a company with different priorities won't acquire us? Could we have known that the response to the measurement requirements of Hoshin would be less than positive, particularly if we'd already implemented a Six Sigma program? How about how different the Hoshin plan started to look when the new CEO came on board? And who could have predicted the Hoshin breakthrough would be implemented in half the time expected, because of a game-changing innovation made in how our industry manages its technology? A reliable practice of Hoshin stays on its toes. Flexibility, alertness, pragmatism, and thoughtful hardworking leadership are the themes of this book.

Organization and Use of the Book

Each article in the book is designed to stand alone, and the articles can be read in any order. Articles 1 through 4 discuss structural components of Hoshin Planning: overall process, organizing for implementation, strategy formation and preplanning, and measurement. Articles 5 and 6 discuss "softer" issues: Hoshin leadership and managing the disruptive transitions provoked by Hoshin. Articles 7 and 8 describe how Hoshin can be applied under two particular organizational conditions: during an acquisition and when using lessons from how startup companies plan.

Article 7 discusses Hoshin Planning that is interrupted by an M&A (mergers and acquisitions) event, and my effort to examine Hoshin during these times moves through somewhat untested written ground. In 2012 and 2013, three of my Hoshin Planning clients were either acquired or made an acquisition. Other than reports using aggregated survey data from big consulting houses such as Deloitte and Ernst & Young, very little is shared from the inside of these events, for competitive and legal reasons. There is even less written about what happens specifically to Hoshin Planning during M&A. My intention in Article 7 is to offer practical advice about how to be proactive and positive, based on my own consulting work, while acknowledging that M&A is generally secretive, disruptive, often exciting, and usually challenging to manage. Most people in an organization will have absolutely no control over how an M&A event goes down. If I err in tone, it will be on the side of seeming excessively hopeful, to balance out what I know to be the difficulty and loss that can attend M&A events. My goal is to open a dialogue, share my own observations and a few from client company executives, and make a contribution to the discussion of Hoshin Planning in an acquisition-intensive economy.

Article 8 on startups is written a little more personally, in part because linking how startups plan to how Hoshin planners in traditional companies plan isn't something I've found much in the work of others. Our Business Centered Learning, LLC office is currently located in the Cambridge Innovation Center in Cambridge, Massachusetts, and we are surrounded by startups. My work and conversations with founders over the last several years have me appreciating what makes their genre of organization special, and my work with client executives in traditional companies practicing Hoshin has given me a receptive playing field for testing some ideas I get from my startup founder colleagues.

Finally, there is an Appendix explaining the general format and use of the basic Hoshin Planning tools referred to in various places in the book: Plan–Do–Check–Act, Affinity Diagram, Interrelationship Digraph, Radar Chart, Annual Plan Table, Implementation Table, Review Table, Catchball, and the Pareto and Run Charts and Fishbone Diagram. Several other books, such as *The Memory Jogger™ II*, cover the construction and use of these tools more thoroughly. The Appendix will simply describe when and for what purpose the tool is used in the Hoshin cycle and provides a real example.

Good Things That Are Not in This Book

There are other topics related to the practice of Hoshin Planning that are important and that interest me, such as the difficulties in moving from planning into implementation; leveraging cross-industry learning; and building a strong weave between the program threads of process improvement, project management, and Hoshin Planning programs in organizations. Those may show up in a future publication.

Also, to discuss the implementation of Hoshin responsibly, we have to consider other well-developed areas of management science, such as change management, project management, organizational design, performance improvement, and leadership development. For the most part, my references to those areas don't include a thorough explanation of the well-known and helpful program frameworks that others have contributed.

Those frameworks give us something to lean on when we talk about the sometimes unpredictable order and messy emotional landscape that accompanies the change associated with Hoshin Planning. I highlight specific practices to try in here, but if you are looking for an organizing guide on these supporting methodologies, please check out the bibliography at the end of the book and the references at the end of each article.

Finally, there are probably fewer standard labels, such as "Communication." Rather, ideas, stories, and suggestions related to key communication activities that make or break the Hoshin are embedded throughout the articles with headings like "Making a Compelling Case for Hoshin as Survival," "Asking for and Receiving Input," and "Describing the Post-Hoshin World." The same is true for "Leadership," which does have an article of its own, but that article does not offer a leadership "model." Rather, the leadership article highlights practices I know work from my time in the field. The book as a whole is written for leaders, and specific guidance for leaders is integrated throughout the articles and delivered through headings such as "Constancy of Purpose," "Speaking with One Voice," and "Having Enough Confidence to Share the Development of the Plan." My suggestions are not exhaustive, though I believe that following some of them will support a better implementation of Hoshin Planning.

Less-Than Hoshin

As the decades-old body of knowledge in Hoshin Planning has been draped over modern environments, innovations and efficiencies in applying the system have been realized. Though an application of Hoshin that strays from the traditional format leaves some people feeling like sneaks. The senior vice president for strategic planning at a client hospital system told me, half-joking, "I feel guilty! Our organization has not embraced Hoshin, with a capital H, per se, except in the quality department. We are more organic. We really haven't taken it down to the 1.1.1.1.1 level of the tables to implement. [pause] What is it about us that doesn't want to be exact about it?"

But his and other organizations *are* disciplined, if inexact, in their practice of Hoshin-*like* planning. The pro forma or lockstep use of the tools and mechanics of Hoshin Planning doesn't necessarily make a planning system better. An appreciation for the Hoshin principles of focus, breakthrough, continuous improvement, change, and inclusiveness might, though. In writing this book, I included Hoshin-like planners in the dialogue, as their input adds value, too.

Stories

This book has references to a few of my closest clients. Focusing on real stories and examples has been a natural choice, as it's through working with sustaining practitioners of Hoshin that my thoughts about the methodology have formed. These executives, plant managers, engineers, physicians, and others who have invited me into their Hoshin Planning work are making the professional world around them better. Naturally, I believe that these excellent individuals, groups, and organizations are the *best*, and I want you to meet them.

Reflection

I imagine these topics look different to different people. They look a little different to me, too, year after year. Writing a book has been an exercise in ongoing reflection. Every time I sit with a client I learn something new, often subtle, now and then profound, and I continue my work in Hoshin Planning at once aware of certain predictable patterns and yet open to changing my mind.

There's an ordinary kind of freedom that I like and that I believe most people working their vocation like: the freedom to think and work at something meaningful and in ways that are satisfying to one's nature and needs. I think this is important, particularly in the application of Hoshin, which demands a higher degree of transformational learning than other workaday practices. We have to like what we do to be doing this hard work. In Spanish there is a lovely way of asking about one's vocation, "A qué te dedica?" What do you do for a living, or literally, to what are you dedicated? I'm dedicated to supporting leaders in bringing focus, good business judgment, and improved quality to the important work they and their organizations do.

Article 1

Hoshin Planning Overview

Contents

> CEO Hoshins are "breakthrough" goals. They are complex issues that usually occur in many of our [Hewlett-Packard's] businesses and often take several years to address. If we can sustain the kind of overall effort that HP people have given on the CEO Hoshins and Business Fundamentals in 1996 as we have done in the last three years, our future is limitless.
>
> **Lew Platt**
> *Former CEO of Hewlett-Packard*[1]

Reasons for Hoshin and Its Role in Planning

Hoshin Planning is a disciplined method for developing and implementing strategic-level improvements in an organization. Organizational leaders increase their success in realizing business objectives with a focused selection of priorities, involvement from a variety of people from across the organization, the use of group-based planning and improvement tools, and the application of a rigorous review process.

The role of strategic plans is to inform our daily actions and decisions over the course of a year or longer. A strategic plan that is referred to and adjusted regularly helps keep the work people do day-to-day relevant to the mission and longer-term objectives of the organization. Some plans achieve this kind of active use, but many suffer from at least three damaging practices.

One: Too Many Priorities

Executive-level strategic plans often try to do too much. The usual eight to ten major changes, additions, or improvements launch out like snowballs from the top of the organization. They roll down the ranks, accumulating heft, until they land on the frontline person as an avalanche of requests for action.

The eight to ten are often called strategic "initiatives." Strictly speaking, initiatives are things that will be *initiated* this year. Instead, Hoshin Planning invites a leadership team to consider what can be *completed* in the next year or two. "Complete" might be defined here as being done to a degree of quality and sustainability that can be operationalized and that will bring the organization to its next level of performance.

Organizations typically don't have the resources to accomplish, to a high degree of quality and completeness, the eight to ten initiatives often put in place. So the strategic plan document becomes a kind of onerous reminder of all that we must, but realistically cannot, get done.

Two: Insufficient Detail

Strategic initiatives at an executive level have a way of looking very reasonable and doable. They show up on presentation slides as neat bullet items. Executive teams may feel some satisfaction: "Look at all we will do this year." Boards may feel they have done their duty by overseeing such a fine list of important activities.

The lack of implementation specificity in the bulleted items, however, though appropriate at the executive level, does not provide sufficient leadership to the organization. Executives often leave out an important transitional step when delivering priorities to their managers.

Only by overseeing the breaking down of high-level goals into smaller pieces can leaders assess the real scope of each effort and its impact on organizational resources. Without some attention to detail, we are left with an understanding of *what* will get done, but missing a thorough idea of *how*.

Three: Lack of Active Review

Strategic plans often lose momentum along the way of their implementation. In an unfortunate twist of organizational physics, many plans develop a kind of inertia. Even motivated leaders who hold parts of the plans can get stuck, if they aren't part of a periodic and rigorous check-in. More innovative, less familiar, or slower-to-start parts of the plan sometimes never really get off the ground.

When we do get around to reviewing strategic plans, we can be sloppy, allowing quick status reports to stand in for real assessment of measurable progress. Without meaningful measures that are tracked regularly, activities that are off-track don't get redirected, successes may not be acknowledged, and strategic plan components that are completed may not get retired to allow contributors to move on to other things.

Hoshin Characteristics

Hoshin = Shining arrow pointing direction. Compass.
Kanri = Control or management.
A Hoshin = One breakthrough objective that the organization will accomplish over the next twelve to eighteen months.

The initial steps of Hoshin Planning look a great deal like any other strategic planning process. Executive teams consider data on business and operational performance and identify current and future customers and their needs. Strengths, weaknesses, opportunities, and threats are considered. The organization's mission and values are reviewed and refined, and attention is given to making sure the organization's measures of success are relevant and sufficient to assess progress on important organizational goals.

Special Features of Hoshin Planning

Hoshin-style strategic planning then departs from this usual path in four important ways: the degree of focus and prioritization, the use of specialized planning and process improvement tools, the involvement of all levels of the organization in defining the plans, and the rigor of the reviews. Figure 1.1

Figure 1.1 Hoshin Planning phases and actions.

shows the typical flow of work for completing a Hoshin Planning cycle. It is through this work that the four special features of Hoshin are made real.

Focus—Selecting "One" Hoshin

The likelihood of success for the strategic plan can be increased significantly by agreeing on priorities at the executive level. This often seems to be a difficult challenge for leadership groups. Temporarily tabling an important initiative can feel irresponsible. Executives can ease the tension around this by staging the Hoshins that are *not* chosen for this year in something like a Gantt chart, assuring teams, boards of directors, and each other that a Hoshin not chosen this year may be in queue for a subsequent year (Figure 1.2).

A Hoshin is a breakthrough, not an incremental change. It's the implementation of an improvement objective that has at least three characteristics. First, it likely requires a fundamental change in the systems of the organization. Second, it requires the support of the entire organization to achieve.

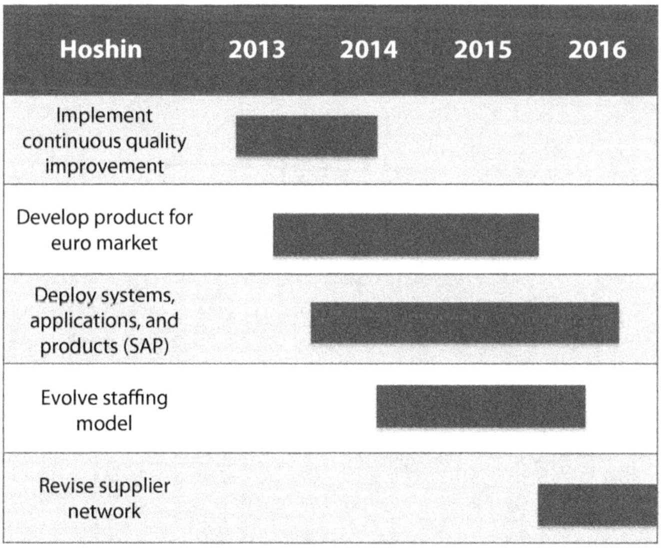

Figure 1.2 Hoshin Gantt example.

Third, if accomplished, it will bring the organization to its next level of performance. To accomplish a Hoshin is transformative to an organization.

Using the Planning Tools and Tables

Detailed documentation of the plan helps enable the level of rigor in review for which Hoshin Planning is known. Hoshin Planning accomplishes much of its early documentation in real time, through the use of group processing tools such as the Affinity Diagram, the Interrelationship Digraph (ID), and the Hoshin Planning tables. A further explanation of these tools is provided in the Appendix. Planning meetings use these and other group processing tools as a supplement to, or replacement of, traditional meeting minutes. The products of the group's work at a meeting become the components of the plan itself.

A leadership team might invest a few hours in the dialogue that produces these charts. How much time it takes depends on the size of the group and the skills of the members as a team. Merck's Vaccine Product Technical Operations (VAPTO) leadership team used the Interrelationship Digraph shown in Figure 1.3 as part of the documentation used to communicate the initial Hoshin Planning work to those outside the leadership team. One of the items shown to influence the most other items in VAPTO's ID is to *Build a Strategic Technical Network to Provide*

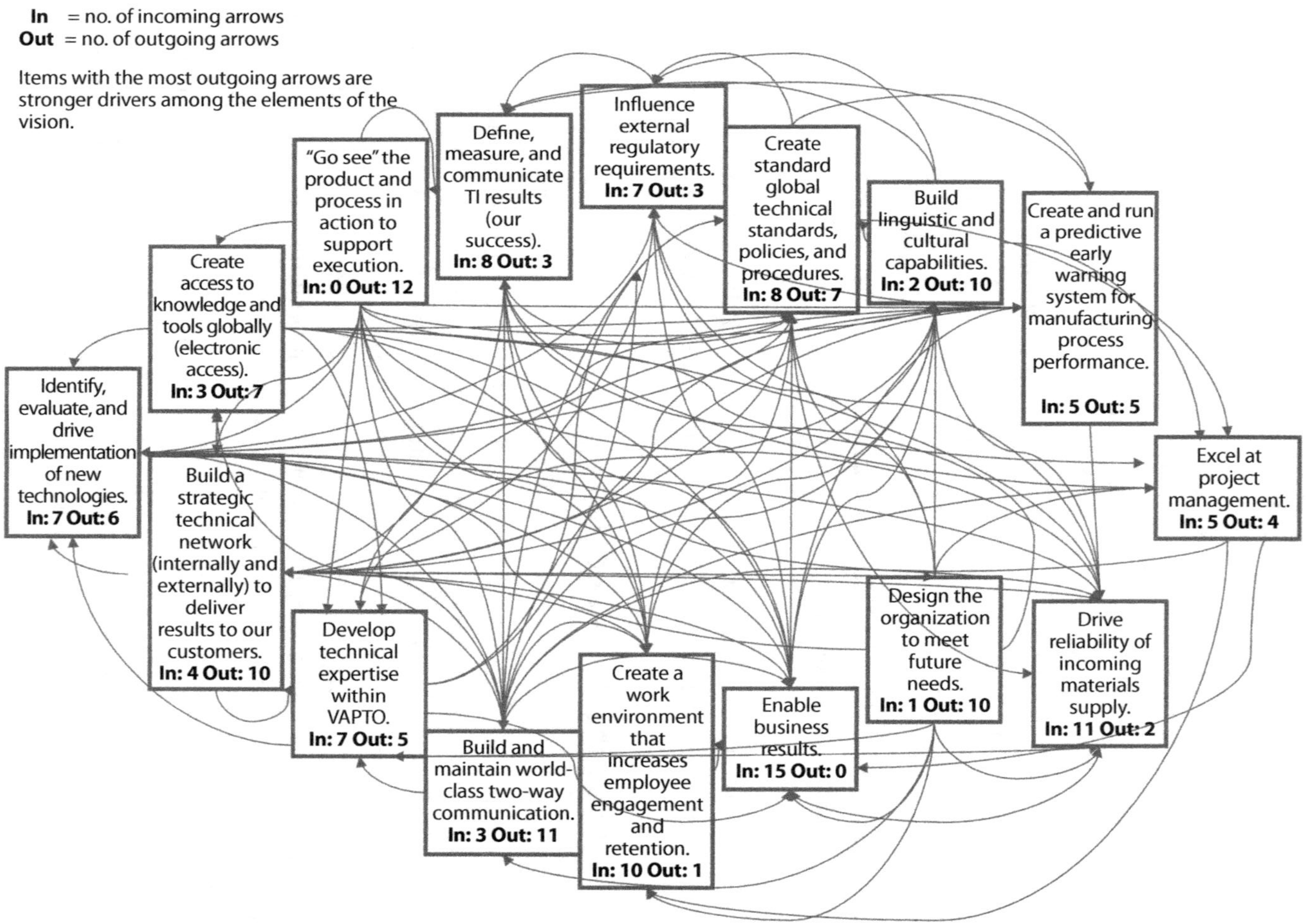

Figure 1.3 Interrelationship Digraph—Merck VAPTO.

"One-Stop Shop" Services. This item was chosen as their annual Hoshin objective.

Tools such as the ID invite a high degree of interaction among leaders and are designed to support consensus decision making. It's a wild looking picture when it's done, but the leaders who work hard with their colleagues to produce these shared images are generally well invested in supporting the Hoshin objective later.

One of Hoshin Planning's core tables, the Annual Plan Table, is used to show the strategies that were used to accomplish VAPTO's *Building a Strategic Technical Network* Hoshin objective. The Annual Plan Table is shown in Figure 1.4. Today, planning documents like this table are often managed with the use of software. Electronic media can make it easier for managers to track progress on plans and make corrections when things run off course. At the same time, real care has to be taken that the exchange of files is not used as a substitute for the important dialogue that advances and accelerates the progress on the plan.

Organizational Involvement—Conducting Catchball

Once the Hoshin is chosen and the highest level of the plan is formed, the leadership team uses an organized process for testing the plan with managers at the next level. The Hoshin practice of Catchball helps build the bridge between the high-level Hoshin plan and its deployment and implementation. This is where leaders seriously consider what it will take to get the job done right.

In Catchball, managers, content experts, and maybe other relevant stakeholders review and provide input to the strategies and metrics that will support the accomplishment of the Hoshin. The process is completed first at the highest level of the plan. Once this review is complete, the strategies are assigned to people who form teams and use the Affinity Diagram again to answer the question, "What are all the things we need to do to accomplish ______?" inserting in the blank the strategy they have been assigned. Their answers to the question form a next level Annual Plan Table.

Imagine completing the exact same exercise again, one more level of detail down in the plan as shown conceptually in Figure 1.5. Soon it becomes clear what volume and type of work will be needed to achieve the Hoshin. For strategies that are unprecedented in the organization, teams may look outside for a realistic assessment of what it will take to accomplish a

High-Level Annual Plan Table

Date: 2012 By: Tech Integration LT **Hoshin Objective: 2012 –** Build, use, and refresh a strategic technical network (internally and externally) to help provide 'One Stop Shop'

Hoshin or Strategy	Strategies	Owner	Measures	Goals
To provide "one stop shop" solutions to our customers, we will build, use, and refresh a strategic technical network (internally and externally).	1. Build the technical network	F.G	% customers using tool	≥ 80%
	2. Define what our customers can count on Technology Integration to deliver	J.P.	Complete VOC*	4Q 2012
Measures	3. Master/use Info Tech solutions for collaboration	S.H.	- Inventory existing tools - Train on tools	- 2Q 2013 - 4Q 2013
Project Efficiency gains (FTE Savings or Execution Speed) # Referrals/ Return customers Customer Satisfaction (measured by annual or per-project Voice of Customer*) External & internal network participation and influence on requirements	4. Develop standard project management processes & tools	B.X.	Inventory of existing tools Gap analysis	2Q 2013
	5. Branding – market & communicate who TI is and successes	M.C.	% TI represented on web page	100% by Q1 2014
	6. Build internal technical skills	J.W.	- # speaking engagements - # outside classes	TBD
Goals	7. Encourage the right behaviors through reward & recognition	J.M.	Criteria for rewards Identification of rewards	Q2 2013
Project Execution time ↓ __% Network 80% populated by 2013 80% rate projects "5" on 1-5	8. Monitor and report our performance measures	A.S.	% work captured Customer Likert 1-5 ratings	100% work captured 100% customers rate 4 or 5
	9. Conduct regular "Grand Rounds" – peer review sessions	J.B.	# of sessions	2 per quarter

Figure 1.4 Annual Plan Table—Merck VAPTO. (Reprinted with permission.)

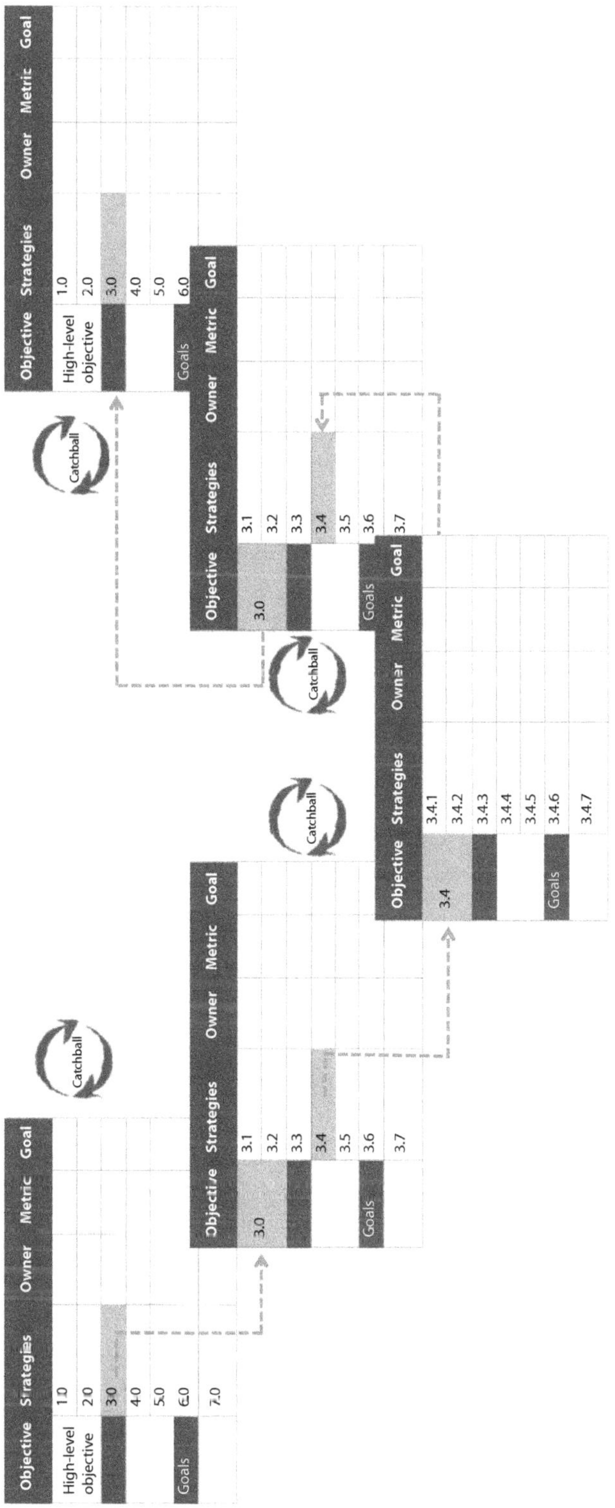

Figure 1.5 Catchball.

strategy. Looking outside can also help assess whether the goals in the plan are realistic.

These more detailed and revised plans are then delivered back up to leadership, and adjustments are made to the plans at that higher level. It's the tossing down and tossing back up of these plans to which the word "Catchball" refers. Often executives are surprised at the resources needed when a vetted set of planning tables is presented. The full appreciation of the resource impact can aid the tough leadership dialogue related to prioritizing the organization's strategic initiatives.

Reviewing the Plans

One of the most impactful features of the Hoshin Planning process is the review. More than the typical progress report presentation, Hoshin Planning strategy owners must report regularly on whether metrics were made or missed, and if missed, what the root cause was that led to the miss. The expectation is that plan leaders will present data and their plans for adjustment as part of the review. "Missed" metrics may be an opportunity for fresh thinking or a clue that a current piece of the plan isn't going to get the desired results.

Care is taken to ensure that review meetings aren't punitive events. Mistakes are used as learning opportunities. Experimentation is encouraged, and risk-taking is rewarded. Well-designed and facilitated review meetings share some features. Good meetings:

- Allow the plan to be iterative, stay flexible, but not diverge from the overall Hoshin objective.
- Focus on measureable outcomes and deliverables.
- Regularly include reflection on the planning process itself, not just its outputs.
- Feature standard reporting tools and practices.

Efficient, focused review meetings held weekly, monthly, and quarterly remind the organization of what is important, refocus derailed components of the plan, and retire parts of the plan that have been completed. Completing reviews is essential. They represent the "Check" in the Plan–Do–Check–Act cycle that Hoshin Planning accomplishes at the enterprise level.

Challenges to Implementing Hoshin Planning

Hoshin Planning presents its own challenges for leaders. Difficulties in prioritizing, no central leader of the Hoshin plan, underestimating the stress on the organization, an organizational deficiency in process improvement skills, unwillingness to change current measures of success, and poor plan documentation practices can all drag a potentially successful Hoshin plan down. Leaders ought to include some mitigation of these challenges early in the Hoshin Planning process.

Reluctance to "Let Go"

It is many times more difficult to remove something from the organizational activity list than it is to put something on it. Projects that no longer contribute to the long-term strategy can be hard to discontinue, but it's essential to do so to make room for work that is better aligned to where the organization has decided it's heading.

No Central Management

Owners of parts of the plan need both administrative and facilitation leadership to keep their respective teams linked to the overall plan. Working on the Hoshin is extra work for most participants, and providing designated, central program management increases the likelihood of success and can reduce unnecessary stress on those working the plans. Leadership teams need to assign an overall Hoshin program manager who is devoted full or near-full time through the full life cycle of a Hoshin implementation.

Underestimated Stress on the Organization and the Resistance to Change

Excellent tools and methods are available for managing the anxiety and dysfunction that often result from introducing a fundamental change in the systems that make an organization run. Leaders must make resources available for what are sometimes called the "soft" skills and include culture and change issues in their implementation plans. A robust change management plan, led at the executive level, is needed to bring about success with Hoshin.

Weak or Isolated Skills in Process Improvement

Just one or two people who know how to complete a good root cause analysis or flowchart a process is not generally sufficient to support the correction of issues that arise during the implementation of a Hoshin. Organizations should consider conducting real-time training in process improvement tools throughout the Hoshin Planning process.

Not Changing Current Measures of Success

If the Hoshin is unprecedented in the organization, leadership may not initially know how to measure it. Overattachment to measures that have been around a long time but are no longer connected to the direction of the organization can distort how the executive team evaluates the success of the Hoshin plan.

Weak Documentation

There can be technical and cultural challenges to maintaining the documentation of the plan. The standard tables related to Hoshin Planning should be linked, centrally accessible, and kept current. Information empowers and provides energy, and moving reliable information around enhances Hoshin's success.

Hoshin Planning

In most organizations, some kind of structured strategic planning is necessary to advance the organization's objectives. In a time when the pace of change is fast and the degree of innovation in technology is high, that planning has to be agile, able to sustain a long-term focus, and also allow for adjustment. Hoshin Planning responds to these conditions well.

Note: Article 1 was adapted from Boisvert, Lisa. "Strategic Planning Using Hoshin Kanri." *Executive White Paper Series.* Salem, NH: GOAL/QPC, 2011.

Endnote

1. Platt, Lewis. "Letter from Lew Platt." *Measure*, November–December, 1995, p. 28.

Article 2

Organizing for Hoshin

Contents

> As objectives cascade down into the organization into strategies, it is prudent to track the manpower required for appropriate implementation.[1]
>
> **Yoji Akao**

A Hoshin plan is a soulless set of tables and charts without the people to usher it through to reality. Typically, when an organization takes on Hoshin Planning, it already has an organizational structure and shared practices in place, with reporting relationships, performance review systems, and the other parts of the Human Resource infrastructure we've learned to expect in the modern workplace. Hoshin Planning overlays a temporary suprastructure, which must respect, leverage, and sometimes work around what is already there. Figure 2.1 shows where in the Hoshin process leaders might pay particular attention to organization.

Figure 2.1 Hoshin Planning phases and actions—Organizing.

Hoshin Roles

There are some existing conventions around organizing for Hoshin—roles, teams, program management, and meeting practices that support the completion of a Hoshin Planning cycle. The implementation of the Hoshin plan often happens through the formation and management of special teams, each of which takes a piece of the Hoshin plan and does the hard work to break that section of the plan into further detail, innovate ways of creating the deliverables it is assigned, and build bridges from the creation of the plan to its implementation.

Each strategy team has a leader and maybe some adjunct members who bring expertise as needed, and each strategy team reports to a program manager, who is the caretaker of the overall plan and ensures that the strategies coordinate well with each other in achieving the overall Hoshin objective.

Hoshin Planning Roles

Strategy Owner
Strategy Team Members
Hoshin Program Manager
Content Experts
Executive Leadership Team

Strategy Owner

Once the Hoshin is selected, and the strategies to achieve it are defined, the leadership team will assign leaders who form teams around the strategies. One person should be assigned to facilitate the work on each Hoshin strategy. That person is the owner of the outcomes of that strategy. Included in that person's job are the following activities:

- Forming and managing the team to achieve the strategy's objectives
- Contributing to the overall management of the Hoshin
- Setting the agenda for and leading strategy team meetings
- Conducting regular reviews of progress on metrics for the strategy
- Representing the work of the strategy team to Leadership and to the rest of the organization
- Following up with people regarding the completion of assignments
- Making connections as needed at her/his own level in the organization
- Teaching the planning process and tools, as needed
- Accepting ultimate accountability for achieving the metrics set for the strategy

The strategy owner positions are critical to the success of the Hoshin plan. It's good to assign "A Team" people to lead the Hoshin strategies. Excitement and investment in the work, from every strategy owner, will help avoid having a short leg on the overall table holding up the Hoshin work. Other strategy teams can struggle when one team falls short in its commitment or its ability to innovate.

Strategy Teams and Members

There is benefit to allowing time for these teams to form and develop connections among themselves. A charter or A3 chart can be a support to this network of teams as they start to come together to define the boundaries between their respective areas of work.

Chartering

There is a large body of excellent work on team formation and chartering. It's mentioned briefly here so that it isn't overlooked as an important step in getting the Hoshin teams going. Figure 2.2 shows an example of a Hoshin strategy team's A3 chart, the chartering template often favored by the Lean and Six Sigma communities. Something like this will help people discuss and agree on the scope and nature of their work together. That clarity is important in a Hoshin Planning context, where what the team is asked to do may be new in the organization and difficult. Some time to discuss and define the work allows for a shared understanding of priorities and maybe some bonding among team members, the connective tissue that can hold people together when the usual challenges associated with large-scale change arise.

Finding High Potentials

So who do you want on the Hoshin strategy teams? These should be the organization's top performers, the hardworking "A Team" players, the most innovative thinkers, the energizers and rainmakers, the healers and consensus builders, the high-potentials—yes, the people who are already much too busy. Because there is no more important or high-stakes work than that which will take the organization to its next level of performance, Hoshin Planning deployment and implementation teams should be populated with the organization's best.

And when strategy owners are assigned a piece of the plan, the dash for HIgh POtentials (HIPOs) begins. Strategy owners often employ a kind of horse-trading when building their teams, swapping favors and wrestling for the best team members. When inviting or assigning people to join a Hoshin strategy team, consider reaching out to people who:

- Have a passion for the Hoshin
- Possess underrepresented expertise related to the Hoshin

CHARTER A3	Title – Strategy Team 8. Monitor and report our performance measures.

Charter Approved Date & By		Date Updated	

Project Description

Develop and implement a set of metrics and reporting tools to report the aggregate business impact of the Department. Collect VOC (Voice of the Customer) to define what metrics will be measured. Implement the process to collect data and report out. Perform effectiveness check and make improvements as needed.

Problem and Goal Statement

Problem:
No method to consistently monitor and report aggregate Department performance and impact. Current lack of visibility to senior management and other groups who are potential customers, impeding ability to expand to other areas.
Goal:
Build, pilot and launch reporting tool with chosen metrics by 4Q2013. 2013 TI Accomplishments are captured in the tool.

Business Outcome / Goal / Benefit

Business outcomes:
- Less individual time spent reporting accomplishments because of efficient tool
- Less time spent understanding the data by customers due to regular reporting frequency.

Business benefit :
- Higher visibility to department Accomplishments enabling more department involvement.
- Data-driven justification for department resources.
- Communication method for defining who we are.

Scope Definition

Deliverables:
Metrics defined by Voice of the Customer input and Strategy Team Reporting Tool (Tier Board, Newsletter, Email, etc.)

In Scope:
Business benefits achieved through completed or active projects.
Customer feedback on TI performance
Future projections of business impact.

Out of Scope:
Individual project work already reported using current project management system

High Level Project Plan

Key Activities / Milestones	Start	End
Define and Collect VOC and Baseline Data	Jan	Mar
Analyze VOC and Define the Metrics	Mar	May
Build the Reporting Tool and Rhythm	May	Jul
Pilot and Launch	Jul	Sep
Establish Feedback Loop	Sep	Dec

High Level Resource Estimate (FTE's & Cost)

Functional Area	1Q	2Q	3Q	4Q
Department FTEs	0.2	0.2	0.2	0.2

Sponsorship and key stakeholders

Assumptions, Constraints, Risks

Assumptions – All of our work will be captured by the metrics (measurement of success of the team)
Constraints – Availability of data.
Risks – Data accuracy, metrics capture all of our work. Participation and reporting from the Dept.

Figure 2.2 Strategy team A3.

- Represent an area of the organization that will be essential to the Hoshin's success
- Are generally recognized as leaders, whether through the formal hierarchy or not

A heartening phenomenon occurs when a person who is junior, or seemingly has few skills relevant to the Hoshin strategy, rockets to informal leadership in the Hoshin work. The management scientists Claudio Fernandez-Araoz, Boris Groysberg, and Nitin Nohria hypothesize that there are a few critical leadership characteristics or skills some people natively possess.[2] Those people can be taught other additional content and technical skills, but not all people with content and technical skills can easily learn those critical characteristics essential to leading well.

A HIPO may have those features that are harder to change, that are already resident in the person, like a motivation to engage people, a sense of resolve around a shared purpose, or a passion for understanding new things. The implication for Hoshin Planning teams is that there may not be much time for shepherding someone to an epiphany that the Hoshin is the best step in the right direction, but there is probably enough time to teach people who already believe in the Hoshin and possess the core leadership skills and the technical components to participate successfully on a strategy team. Strategy owners recruiting for their teams can get a big win in what these three scholars call the "war for talent" by relaxing some biases about who might contribute best.

"You want people who are capable of building something from scratch and seeing it through. Finding them frees the rest of your team to work more and manage less," write Fried and Hansson, founders of 37Signals, in their book *Rework*.[3] This entrepreneurial spirit and capacity for follow through is key for creating a group of people around the Hoshin plan who can make it happen.

Hoshin Program Manager (aka Hoshin Hero)

One CEO calls his Hoshin program manager his "Free Safety." In football lingo, that's the player who can handle any position and is a leader on the field. The Hoshin program manager holds all the components of the Hoshin plan together and works closely with executives in keeping the Hoshin plan on track.

With any enterprise-wide and strategically critical plan, if you assign a part time lead, you can expect partially good results. *It is critical to have one central person whose primary job is to keep the high-level Hoshin plan focused and moving.* This person oversees all the strategy teams, maintains the overall planning project schedule, serves as a liaison with the consultants involved in the Hoshin work, and often shares in the facilitation of the executive team's completion of high-level decisions related to the Hoshin.

The thinking behind choosing the person for this role shouldn't necessarily rely on the criterion of seniority. For the right person, it can be an excellent professional growth opportunity, a way to leverage the Hoshin Planning effort in service to the organization's succession plans. It's best to choose someone who is meticulously organized, substantially open-minded, and widely respected.

Co-Leadership—No

If you'd like to put some thought into assigning two leaders to a strategy and test some assumptions about the conditions that might make that work, you might read David Sally's engaging article, "Co-Leadership: Lessons from Republican Rome."[4] Most times, though, assigning a single leader who is accountable for results in his or her sphere of the plan is the way to go.

Content Experts

Finding people with specific expertise related to a Hoshin strategy and bringing them in on an "as needed" basis is a way to get more expertise into the Hoshin work without loading up and slowing down teams with too many members. Those people may be more supportive of the results of the Hoshin work, as they've been offered a meaningful opportunity to inform and shape the outcomes.

Another nice by-product of rotating content experts on and off teams is that by doing so the network of people exposed to the Hoshin expands. These people often head out into the organization and talk about their experience contributing to the team. Their freedom from the daily, weekly, and monthly grind of the Hoshin strategy team's work can produce a lightness in how they talk about the work, important energy to leverage during the Hoshin implementation.

Executive Leadership Team

Of course, the central component of the Hoshin Planning organizational structure is the leadership team itself, the team that conducts the initial environmental and organizational assessment, creates the vision for the future, and chooses the Hoshin. Article 5 in this book considers the formation, importance, and health maintenance of leaders and leadership teams.

Once the Hoshin plan is launched and people at various levels of the organization are contributing, the leadership team's job is one of removing barriers to the teams' work, negotiating among departments at the executive level, reinforcing the scope (what's "in" and what's "out"), moving resources around, and constantly communicating the reasons for the Hoshin work. Some steady management and instruction from the leadership team are necessary to keeping the plans moving in the right direction.

Organizational Structures and Tools That May Support Hoshin

There are tools for allocating accountability and defining reporting relationships that can support Hoshin Planning. Sometimes those tools are employed temporarily and exclusively for the Hoshin work, and sometimes they are adopted as the primary way the organization does all of its work. Four of those are discussed in the following section.

Structures/Tools

- RACI, RAPID, and Other Decision Tools
- Open Allocation
- Lattice
- Matrix

Responsibility and Accountability

Organizational structures are substantially about deploying, managing, and reinforcing responsibility and accountability to get things done. That's a lot of what Hoshin deployment is about, too, and how the Hoshin is deployed should be informed by the structure of the organization. It's somewhat counterintuitive, but real innovative and free thinking seems to be enhanced by clarity about where the lines of accountability lie and what level of authority each person is expected to exercise to advance a shared objective. In his classic and beautiful work *Escape from Freedom*, Eric Fromm suggested that some decentralized and logical structure of accountability is needed to reduce the anxiety that can attend confusion about what role someone is playing in helping a higher goal progress.

The platform of stability provided by clarity on roles and responsibilities can invite freethinking and creativity. Fromm wrote that to fully experience a democratic leadership toward a common set of priorities, every individual must "share responsibility and use creative intelligence in his work."[5] Accountability must be spread around and down into the organization to allow for local decision making and creativity. On this Fromm writes, "The active control and co-operation by each individual and by the smallest units of the whole system requires a great amount of decentralization."[6] Giving people agency over their work generates energy and invites deeper commitment to the organization's goals.

There are independent tools that can outline and express this sharing of accountability, regardless of the existing organizational structure. One of those tools is a chart, sometimes called RACI and sometimes ARCI, which helps define for each decision associated with the Hoshin work whether a position or person is "Responsible, Accountable, Consulted, or Informed" relative to that decision. Figures 2.3 and 2.4 show a common structure for applying this thinking.

Responsible	Responsible for performing an activity. There can be more than one *R*. *R* implies *C*.
Accountable	Ultimately accountable for the outcome. Includes yes/no and power of veto. Only one *A* can be assigned to an activity or decision.
Consulted	Consulted prior to a final decision or action being taken. Involves two-way communication.
Informed	Informed after a decision or action is taken. One-way communication.

Figure 2.3 RACI definitions.

	Position or Group	Position or Group	Position or Group	Position or Group
Activity	*A*	*R*	*C*	
Activity	*R*	*A*		*I*
Activity	*A*	*C*	*R*	*R*
Activity	*C*	*C*	*A*	*I*

Figure 2.4 RACI conceptual chart.

Hoshin executives and program managers can use tools like this to stimulate important discussions around how decisions will be made in the Hoshin Planning work. Clarifying decision-making practices is particularly important in the context of Hoshin Planning, where the usual lines of authority may become blurred or intentionally renegotiated to allow for flexibility in how strategy teams complete their work.

There are other similar decision-making frameworks, such as RAPID, that support this disambiguation. The acronym Rogers and Blenko offer stands for Recommend, Agree, Decide, Input, and Perform.[7] People assume one of these five roles in relation to a decision. Like RACI, the RAPID tool can force the discussions around decisions that need to happen as the Hoshin organization forms and starts to interact with the existing system for managing accountability in the organization.

Open Allocation

Some organizations use what is called "open allocation," a way of organizing around work that allows for more self-determination. For Hoshin Planning, this might mean that teams add members as needed, and members actively retire from a team when their work is done. Strategy owners form the teams they want, by making compelling invitations to the people *they* consider to be HIPOs. Open allocation also means that to some extent people can self-select themselves onto a Hoshin Planning team. Individuals then in a sense "apply" for inclusion on a team doing work for which they have a passion.

One of the exciting times in the Hoshin process is right after the strategies have been defined and the strategy owner chosen. Those strategy owners are welcomed into the leadership meeting and given the mandate in a formal way. Allowing each strategy owner to form the initial message around a strategy and attract the people he or she wants can build a sense of agency and confidence for the strategy owner. This contributes to building teams that are vibrant and functional. If assignments of people are then made to the strategy team and those people assigned don't meet the criteria but are, say, political appointees, a critical mass in the team can be extremely effective at sustaining positive norms and high productivity. A byproduct of their solid team function can be the shaping of an appointee's contribution in a way that stimulates growth for that person and benefits both the Hoshin work and the appointee him- or herself. It's important that the strategy owner be able to form the core of his or her own team.

The communication software innovator GitHub is structured in what it calls a "post-bureaucratic" way, allowing a great deal of self-determination for its employees in defining their work. Hoshin Planning, because of its high-stakes nature strategically and the newness it represents to the organization, invites similar thinking. But communication in this open format is critical. GitHub's former CEO, Tom Preston-Warner, described a web of information and self-selection like this: "In some ways, the structure is just a distribution network for information about *the purpose of this company and how it works* and subsequently *what it should be making*. Keeping everyone on the same page about the incremental adjustments in direction is really important—otherwise, word about these small corrections doesn't make it to certain units of the business, which don't adjust, eventually leading to huge gulfs in purpose between teams."[8]

The Lattice

This idea that structure influences how information is moved around is consistent with a commitment to helping people emerge as leaders, so that they can be redeployed to the next Hoshin or other high-stakes parts of the organization. Bill Gore, founder of the company that invented Goretex in the 1960s, described allowing "special teams [to] evolve within the structure, usually led by someone particularly competent in the discipline or activity of the team. Every (successful) organization has a lattice organization that underlies the façade of authoritarian hierarchy," he wrote in the paper "The Lattice Organization: A Philosophy of Enterprise." By lattice, he means the informal substructure of the organization. Gore wrote, "It is through the lattice organizations that things get done, and most of us delight in 'going around' the formal procedures and doing things the straightforward and easy way. In the process, natural leaders will evolve."[9]

A kind of ghost organization is where many things happen in an organization, and it's there that we can gain from experimenting with letting the strategy owners and their teams do their thing, form their own networks, and try things that are not usually done in the organization. The more freedom they have, with clear measures and boundaries defined by leadership, the more invested people may be in the work they're doing.

To the greatest extent possible, it's important to get people on Hoshin strategy teams who *really* want to be there. A strong desire to learn the new thing should be enough to qualify someone for a Hoshin strategy team. Hoshin executives ought to prioritize breakthrough thinking in the Hoshin work and think beyond the obvious people in content positions in the organization. Teams can benefit, too, from the fresh thinking that comes from new people asking what might be seen as amateurish questions, but that invite a new look at old problems.

The Matrix

Excellent work has been done around the benefits and challenges of matrix management. Some version of the matrix is common for structuring the Hoshin deployment. Despite some difficulties, it's a fairly good way to build the temporary organization needed to implement the Hoshin plan. The difficulties may include things such as competition for resources and feelings among department leads that they are losing some control, things that can and should be managed to avoid unhealthy conflict in what the architect/psychologist team Allen and Henn described as the "Battlezone," a conceptual picture of which is shown in Figure 2.5. The D's are departments, the P's are projects. Hoshin strategy owners get in the ring with line managers to negotiate designs, pace, and resources for the new products and processes that will result from the Hoshin work.

The work in this zone has potential to be especially productive. Hoshin teams are often given a hall pass to leave the organizational space they were hired into and play in an open space in the organization. At the same time, authority ought still to be invested in department leads to maintain the

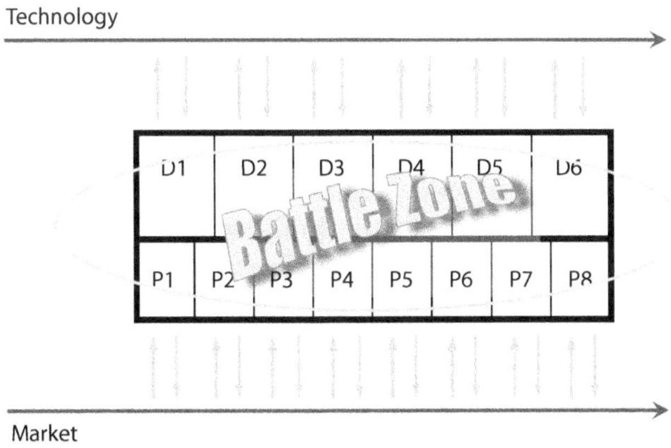

Figure 2.5 "Battlezone." (Reprinted from *The Organization and Architecture of Innovation: Managing the Flow of Technology,* Allen, Thomas J., and Henn, Gunter W., p. 85, Copyright 2007, with permission from Elsevier.)

quality of products and services created through the Hoshin and to ensure time to market. Tolerating the tension between Hoshin strategy leaders and department heads can be helpful to achieving good results.

The tension can also provide a kind of "checks and balances" on key decisions, ensuring an idea or choice is more rigorously vetted before moving forward, the vertical voice often representing cost and sustainability and the horizontal voice representing innovation and experimentation. This can slow things down, but can also surface difficulties earlier in a process. Key to managing the tension is the insistence that both "sides" of the matrix agree on specific metrics that will define the success of the work.

If there isn't an existing matrix organization in place, the temporary suprastructure of Hoshin often puts strategy owners in the position of horizontal managers, working across departments to get key projects associated with the Hoshin plan accomplished.

Leveraging Traditional Human Resource Practices

There are existing mainstream human resource practices that most organizations follow. Hoshin planners can tap these to leverage their activities and to add value back into the organizations. Two of these practices stand out as especially linked to Hoshin Planning: succession planning and performance review.

Succession Planning

Hoshin Planning forces organizations out of their comfort zones and into breakthrough areas that will launch a step change in performance. The workspace created by this kind of change presents an opportunity to expose and invest in new leaders who will pick up and carry the changes over the next several years. "When you're creating something new, something nobody in the company has experience with, and you have people you want to test, they can't just go get someone else's game plan. Seeing people independently demonstrate leadership by creating something new tells you everything you need to know about somebody,"[10] says NV Energy Hoshin Planning executive Kevin Geraghty. Hoshin is a good testing ground for people targeted for potential growth in the organization.

Performance Review

Introducing performance review during a Hoshin Planning meeting is a good way to experience a loss of meeting momentum. Still, some good can come out of adding structure to the expectations we have for each individual in the Hoshin Planning process. Again, people pay attention to what gets measured.

Hoshin leaders sometimes weave the measurable objectives from the Hoshin plan into individual Key Performance Indicators (KPIs) on which an employee is evaluated. The appetite for this kind of hardwiring of the Hoshin measures into performance review varies.

Space as Tool for Organizing Hoshin Work

When a Hoshin objective is implemented, it often involves dialogue with individuals not usually on one's meeting schedule, tapping on doors in other departments, and in general getting into each other's business. Hoshin planners move information around, inspire people to think innovatively about the Hoshin, and organize the work to bring those innovations from planning into action. The people doing these things seem to have an easier job of it if they are located near each other. Considerations such as centralizing conference rooms, maximizing face-to-face meetings whenever possible, and rotating the locations of routine meetings can all help accelerate the pace and improve the quality of the Hoshin work. The architects Allen and Henn write about the effect of proximity on innovation, "Awareness is at the very center of the innovation process. Success … depends on all those involved being aware of the invention or idea and the state of its development."[11] The physical workspace can make that ongoing awareness easier to sustain.

More than emailed status reports are needed to keep the fabric that holds the Hoshin work together strong. The more people see each other, the more likely they are to pick up the phone and call someone on another Hoshin strategy team for information. Figure 2.6 shows the results of a study by Allen and Hauptman correlating face-to-face meetings with telephoning.

In Figure 2.7, Allen and Henn's research further supports the idea that the further away we are from each other, the less we'll communicate. Communication becomes less frequent, less than once a week, when we work as little as fifty meters apart. And the probability of our communicating by phone drops in situations in which we don't regularly bump into someone in person.

Perhaps for the greater degree of innovation usually required in Hoshin Planning more support of what collaborative space advocates call serendipitous collaboration should be considered. Temporarily co-locating Hoshin planners may enable rapid prototyping, creativity, and reduce duplication of efforts. Companies like Zappos and Google are now sending their people out to co-working spaces to stimulate thinking and connections and even inviting strangers from other companies into their own office spaces to encourage idea exchange and push the pace of innovation.[12]

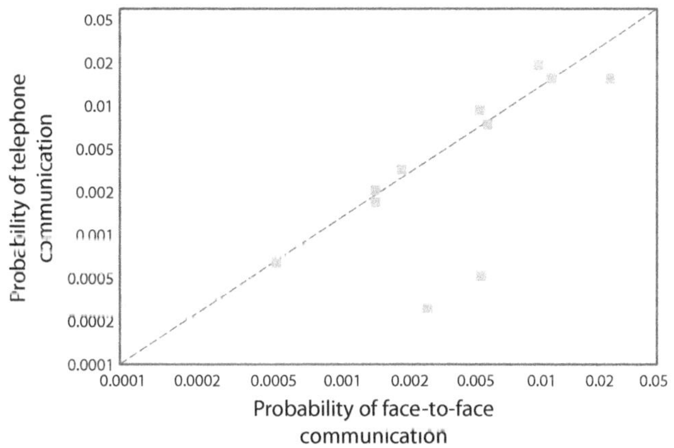

Figure 2.6 Telephone and face-to-face. (Reprinted from *The Organization and Architecture of Innovation: Managing the Flow of Technology,* Allen, Thomas J., and Henn, Gunter W., Copyright 2007, with permission from Elsevier.)

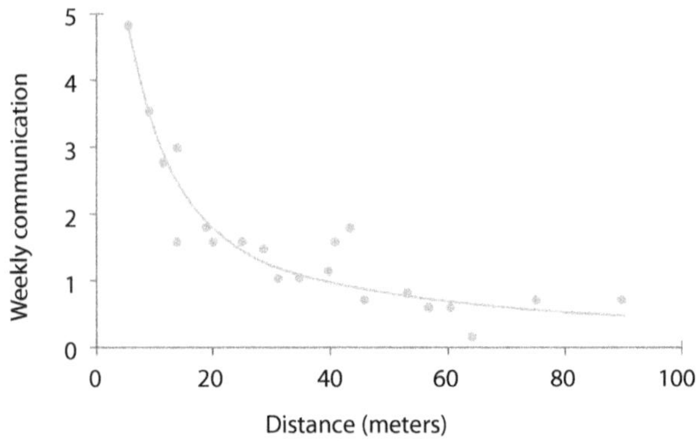

Figure 2.7 Probability of communication. (Reprinted from *The Organization and Architecture of Innovation: Managing the Flow of Technology,* Allen, Thomas J., and Henn, Gunter W., Copyright 2007, with permission from Elsevier.)

Three researchers at Harvard Biomedical Research lab hypothesized that physical proximity produced better collaboration and higher quality in the production of research papers. Lee Brownstein, Isaac Kohane, and R. G. Mills concluded that "Despite the positive impact of emerging communication technologies on scientific research, our results provide striking evidence for the role of physical proximity as a predictor of the impact of collaborations."[13] Leaders may consider temporary measures to get the people working on the Hoshin plan closer to each other as they advance their work.

This may also help avoid a new phenomenon space experts and organizational psychologists are calling "presence disparity," a kind of unfairness in how one's input is valued, depending on whether the person is physically present or contributing via online meeting tools or phone.

Labor Unions

You cannot wait them out, ignore them, or assume that union leadership is prioritizing the dialogue with its constituency through a filter of current business reality. In 2014, between 4 percent and 24 percent of every U.S. state's workers were represented by a labor union,[14] so some leaders will be leading their Hoshin plan with union considerations as part of their work.

Seasoned leaders generally appreciate the value of working hard to bring the union voice into the Hoshin dialogue and are often quite sanguine about the extra challenges. "Some things you can say and some things you can't, when talking with unions. And any time you limit discussion you limit discovery and slow it down. It's just another one of those things you have to work through,"[15] said Kevin Geraghty, Hoshin leader and vice president for energy supply at NV Energy.

Among the special considerations for leaders are how to time and pace the Hoshin work with unions and include the input of union employees in the Hoshin dialogue.

Unions' Impact on Timing, Pace, and Priorities

Letting the union influence how fast or slow you go and what activities move to the front of the line in the Hoshin work can pay off.

"I'd say we considered carefully when to start talking with the union. Our *Workforce Redesign* Hoshin impacted them directly, and the timing of our work on it relative to contract negotiations was important. One success we had was putting together the training strategy that would be used in the new workforce design. It was the number one priority for union members when we surveyed them. That has to be a careful consideration, what's important to the union,"[16] says John Frankovich, Hoshin program manager for NV Energy

Leaders can build a Hoshin Planning calendar that includes and works with important dates for the union—negotiations, litigations, and other union leadership meetings. The planned milestones for the Hoshin should acknowledge, but not overly adjust to the union's preferred pace, which may be slower or faster than what is wise for the business and responsive to industry reality. When prioritizing portions of the Hoshin plan, leaders may want to add extra weight to items that are especially meaningful to union members, if they make sense to the business and the overall Hoshin plan, too. In short, a conscious and agreeable approach to timing and pacing the Hoshin work with union employees can ease the implementation as the changes in the organization become real.

Soliciting Union Input on the Hoshin

There can be a kind of approach–avoidance dance between Hoshin leaders and their union employees as the typical early-in-the-cycle Hoshin Planning step of getting input from senior managers on the plan is initiated. It's conducting "Catchball" in an environment that intentionally builds a filter between the voices of some employees and their leadership.

Here is the language, written by one client's staff attorney and included at the opening of the e-survey sent to all employees to solicit input on the Hoshin objective in its early stage of development. This language was the opening page of the survey, and after reading it survey takers were required to click "Opt In" or "Opt Out" before beginning the survey.

> Nothing in this survey is intended to constitute an attempt to circumvent the collective bargaining process where required. The Company and Union have agreed that the intent of this survey related to [Insert the name of the Hoshin] is to involve a broad representation of employees, including bargaining unit members, in the evaluation process to create a plan for a successful future. It is the intent of this process to have a constructive dialogue where both parties can proactively share concerns and focus on improvements and solutions.
>
> ___Continue with this survey
> ___Opt out of this survey

It may seem a bit awkward to rely on legalese, but for this group it was a practical and effective way of including all employees in the same dialogue. Insisting that everyone, union and nonunion, accept some responsibility for providing input or not is more than a disclaimer that protects leaders later from claims that union employees were not invited into the dialogue. It reflects a conscious choice to get that good input from union employees, too, despite the sometimes difficult path to receiving the input directly and unfiltered. In the case of the company division that used the "Opt In" mechanism in its survey, 265 people completed the survey, and only 7 opted out, a fairly good turnout. It's probably better for leaders to lose a little input than to let the difficulties of collecting the input stop the process of soliciting feedback altogether.

Organization of Hoshin Meetings

The iterative, transformative nature of Hoshin Planning calls for meetings that have enough flexibility to let people talk about unplanned results or topics emerging from the Hoshin work, but that also ensure a neat, predictable structure to make certain the necessary planning work gets done. Balancing the need for structure with the need to stay agile and allow for emergent thinking is needed for all Hoshin meetings—leadership, strategy team, all-hands, and review.

Leadership Meetings

At leadership level Hoshin meetings, attendance should include only those people who will be personally accountable for the results of the implemented plan. In other words, observers, advisors, and other guests can both increase the size of the group in a way that limits the intimacy and depth of the discussion and introduce distractions. Leaders need an opportunity to talk candidly with each other about their concerns and to work efficiently and knowledgeably to resolve issues. Using Hoshin leadership meetings as a "runner up" prize for an individual who was not promoted or wasn't included in another high-profile project or as a kind of "thank you" to individuals who have made a significant contribution to the organization is inviting in some ways, but it's not the way to get a small, high-functioning team of leaders into an efficient and meaningful mode of dialogue around the Hoshin.

By contrast, periodically including strategy owners or key content experts from strategy teams is a good way to keep the leadership team "in the know" on emerging knowledge that is likely to come about from good Hoshin practices. It's also a good support to the organization's succession planning—giving high-performing Hoshin planners some top-level exposure.

Finally, the Hoshin program manager should be included at the highest level Hoshin Planning meetings. That person is managing and coordinating the work of the Hoshin strategy teams according to executive priorities and must hear those priorities firsthand. As mentioned earlier in this article, if the right Hoshin program manager is chosen, that person possesses the discretion and professionalism that will make him or her an asset at leadership meetings.

Strategy Team Meetings

Once a Hoshin objective is selected, several strategies are identified for achieving it. Around each of those strategies a team is formed and meets routinely to complete its work. In addition to routine strategy team meetings, the Hoshin program manager needs to bring all the strategy team leaders together to coordinate the work among them. Figure 2.8 shows an outline of activities, key success factors, and typical outcomes of "all strategies"

Activity	Key Success Factors	Output(s)
Deliver Deliverables	• Only major deliverables from a strategy are presented—ones that impact the work of other Hoshin strategy teams or have immediate impact on the organization • The presenting strategy owners ask specific questions of other strategy owners to guide the approval or rework of the deliverable	• Deliverables that are approved for broader distribution, for example new processes or models or prototypes a strategy was formed to create
Manage Overall Hoshin Program	• Hoshin program manager fully prepares and presents overall milestone schedule and other high-level planning documents • Strategy team members have reviewed high-level planning documents before the meeting and are prepared to comment on the impact on/of their strategy	• Updated high-level plans • Confirmed completion of previous planned changes • Announcements or requests the Hoshin program manager will represent to the Leadership Team
Report Progress on Individual Strategies	• Use of Hoshin planning tables and the development of clear, digestible graphic depictions of the work for broader consumption • Presenters are prepared with explanations and solutions for providing deliverables that were expected, but not delivered	• Changes to the plans • Agreements/requests strategy owners have of each other or the Leadership Team

Figure 2.8 Activities of "all strategies" meetings.

meetings. In the early stages of a Hoshin implementation the "all strategies" meetings might take place monthly. Later in the cycle, quarterly may be enough.

All-Hands Meetings

The angst around who to invite, who will speak, and what can be made "public" can be intense before meetings in which the Hoshin work will be shared with increasing numbers of people in the organization. These meetings should increase in frequency as the Hoshin plan moves into active development and implementation.

Early all-hands meetings are a great place to observe who steps up to informal leadership and how they talk about the Hoshin work. In this way, leaders might allow their choice of Hoshin spokespersons and team members to emerge continually. There is a great deal of change experienced through the Hoshin Planning process, and it's ok to move people around as the conditions change and as you get a better sense of peoples' levels of commitment and skill.

Review Meetings—Decision Making

Hoshin review meetings can be organized to solicit an excellent and productive experience for those involved. There are at least three practices that help leverage the time people spend in review meetings:

Practices for Successful Review Meetings

- Document Standardization
- Meeting Design and Facilitation
- Participation and Structured Decision Making

First, to the greatest degree possible, Hoshin planners ought to use common planning document templates and follow common reporting conventions. Using planning templates shortens the time it takes to present and discuss any particular piece of the plan by eliminating the need to get reoriented with each presentation. In addition, using a common convention such as "made/missed" or "green, yellow, red—good, caution, stop and correct" allows for quick identification of problem areas in the plan, so that meeting time can be focused on those. In general, using a common reporting format, the Review Table, allows those reporting and those listening to focus entirely on the content and not on the means for delivering that content.

To that end, many if not most Hoshin planners today use to varying degrees software to develop and manage their Annual Plan Tables, Implementation Tables, and Review Tables (see the Appendix for more on each of these tools). At the most integrated use of technology for plan management, strategy owners and leaders can tap directly into a cloud- or intranet-based system from within their meeting space, without the use of any paper at all. These systems have the additional benefit of allowing planners to make updates to the plan in real time, allow for up-to-date plan visibility to key leaders and other strategy owners at all times, and produce simple built-in reports without having to move content between spreadsheet and presentation softwares not originally designed for networking and sharing. At the time of this publication, there are two software systems especially made for Hoshin Planning: PlanBase (planbase.com) and WebHoshin (tqe.com).

Next, the design and facilitation of review meetings may be different than for other types of meetings. Review meetings are extremely focused, and they may involve less meandering discussion or debate. It is expected that strategy owners reporting at a review meeting will have done their homework to identify root causes for missed metrics, and that they will have evaluated countermeasures and plan adjustments prior to the meeting. It can take a while for people to get into this focused mode of meeting, and having a skilled facilitator makes a positive difference. Superstars from previous Hoshin Planning cycles can serve well as facilitators.

Finally, skipping Hoshin review meetings should not be an option. By committing to a Hoshin, leaders are saying that this one strategic priority will be the top priority for the next twelve to eighteen months, and when priorities are in conflict, this one will win. Decisions important to the overall Hoshin are made during review meetings, and strategy owners base decisions related to their own team's work on what other strategies are doing. Gaps in representation cause delays, omissions of important information, misinformation, and frustration among other Hoshin planners.

Once people are there, leaders ensure that structured decision-making habits develop and are sustained and that participation is balanced. Article 5 on leadership discusses decision types and tools. Spreading around accountability for the leadership, facilitation, and communication of results to the organization of review meetings can help invest senior leaders evenly in the success and follow through on the Hoshin work. Leaders can model this investment with their teams by taking their turn at leading team maintenance activities.

Some Tactics for Successful Review Meeting

- Standardize documentation practices.
- Ensure representation from every strategy, at every review meeting.
- Maintain a minimum frequency: monthly at first, then quarterly.
- Pay attention to team health and function.
- Pay attention to how time is allocated on the agenda.
- Build in structured decision-making practices and tools.
- Assign facilitation responsibility for every meeting.

Work that happens both before and during review meetings includes aligning activities between strategies, creating and adjusting the overall planning calendar, updating the high-level plan, resolving change management issues, and executive review and revision of charters and plans of teams as needed. All of this work should not be suspended until meetings take place, but advanced between meetings. Between-meeting work is necessary to maintaining the level and pace of work required to accomplish the Hoshin plan.

Organization as a Hoshin Enabler

When Hoshin Planning is described as a disciplined approach for defining and implementing a transformational change in an organization, part of what that adjective "disciplined" refers to is the structure that organizes people together around a shared Hoshin objective. Leaders can strengthen the Hoshin process by choosing an organizing architecture that is compatible with the organization's culture and supportive of the Hoshin Planning process. With too much structure, the innovation required for a breakthrough is stifled. With too little structure, the work lacks the glue to hold it together.

Endnotes

1. Akao, Yoji. *Hoshin Kanri: Policy Deployment for Successful TQM.* Cambridge, MA: Productivity Press, 1991.
2. Fernandez-Araoz, Claudio, Groysberg, Boris, and Nohria, Nitin. "How to Hang on to Your High Potentials." *Harvard Business Review*, October 2011, HBR.org.
3. Fried, Jason, and Heinemeier Hansson, David. *Rework*. New York: Crown Business, 2010.
4. Salley, David. "Co-Leadership: Lessons from Republican Rome." *California Management Review*, 444, Summer 2002, pp. 84–99.
5. Fromm, Eric. *Escape from Freedom*. New York: Holt Paperbacks, 1941, p. 271.
6. Ibid., p. 272.
7. Rogers, Paul, and Blenko, Marcia W. "Who Has the D?" *Harvard Business Review*, January, 2006, pp. 52–61, 131.
8. Dannen, Chris. "Inside GitHub's Super-Lean Management Strategy—And How It Drives Innovation." *Fast Company Digital*, October, 2013, http://www.fastcompany.com/3020181/open-company/inside-githubs-super-lean-management-strategy-and-how-it-drives-innovation.
9. Angrist, Stanley. "Classless Capitalists." *Forbes*, May, 1983, p. 124.

10. Geraghty, Kevin. NV Energy, June 6, 2015, from interview.
11. Allen, Thomas J., and Henn, Gunter W. *The Organization and Architecture of Innovation: Managing the Flow of Technology*. Burlington, MA: Elsevier, 2007.
12. Greg Lindsay. "Coworking Spaces from Grind to GRid70 Help Employees Work Beyond the Cube." *Fast Company*, March 2013, http://www.fastcompany.com/3004915/coworking-nextspace.
13. Lee, K., Brownstein, J. S., Mills, R. G., and Kohane, I. S. "Does Collocation Inform the Impact of Collaboration?" *PLoS ONE*, 5(12): e14279, 2010. doi: 10.1371/journal.pone.0014279.
14. Bureau of Labor Statistics, U.S. Department of Labor. "Union Membership by State in 2014." *The Economics Daily*, http://www.bls.gov/opub/ted/2015/union-membership-by-state-2014.htm.
15. Geraghty, Kevin. Ibid.
16. Frankovich, John. NV Energy, June 20, 2014, from interview.

Article 3

Strategy and Hoshin Preplanning

Contents

> **Walking on a high wire is inherently dangerous, but you've said that you prepare so much that death is not a risk.**
>
> No, it's not. I think this is very wrong to put your life in danger or to gamble with your life. I cannot do the first step not being absolutely sure that I will successfully perform the last.
>
> **Phillippe Petit**
> *May 11, 2014 New York Times Magazine Interview*

As a methodology that invites breakthrough thinking, Hoshin can support leadership teams in walking out onto a market high wire, where things are exciting but the fall can be tough. Anticipating what customers and markets might do, and then using those predictions to choose a market strategy, involves some tolerance for risk. To mitigate and balance that risk, Hoshin Planning practitioners work to have good market information; sound assumptions about the regulatory, competitive, supplier, and political environment; and an honest understanding of their own organizations' gifts and shortcomings. The "life" of an organization using Hoshin Planning depends not only on the ideas and vision of a leader, but also on the due diligence applied to preplanning thinking. Hoshin Planning does not offer leaders a way to be absolutely sure, but it does create a measureable and systematic pathway for making an organization's market strategy real.

It's a tricky balance, to be both predictive about the future and agile enough to allow for a fresh, unexpected strategy to emerge part way through a Hoshin Planning cycle. Strategy makers must adjust as customer behaviors shift, technology changes, and the environment evolves. Perfect prescience isn't as important as leaders intimately understanding the field on which they are playing and staying committed to a planning system that responds quickly to new information. Risk is an expected feature of pursuing a breakthrough with Hoshin Planning, and good preparation is one way to not only reconcile the risk, but also leverage the benefits that result from taking it.

Hoshin Planning ≠ Strategy

Consideration of a market strategy is its own piece of work, one that ought to precede and accompany the Hoshin Planning process. Once the strategy starts to form, Hoshin helps prioritize, align, deploy, and review the activities that move the organization into that strategy. A clear and viable value proposition and defined market position make an important difference to the quality of the results the Hoshin plan can deliver. Hoshin will help focus and apply resources to the work that will take the company to, and keep it in, a chosen market position. Hoshin is a strategic-level tool for operational effectiveness; it does not define market position, or market strategy. It does provide strategies for gaining and keeping the chosen market position.

It's important that someone from the executive office presents the company's market strategy with some specificity to Hoshin planners at other levels in the organization. Teams that create the detailed plans for implementing Hoshin objectives take different approaches and bring different ideas to their work depending on what seat the company is working to take in the industry.

If the strategy is not fully formed at the executive level, there are a number of ways to develop it. This article dips into three general sources of market strategy: frameworks, Hoshin review, and visionary CEOs.

Some Sources of Strategy

Strategy can come from different places, and its source can influence the initial steps of the Hoshin Planning process. For example, a strategy formed with the help of a strategic framework can stimulate detailed ideas in the initial Hoshin Planning phase of creating the "Vision Elements" (see the Appendix for the Vision Affinity Diagram). In general, journeys to the strategy that have involved difficult, high-quality, and option-limiting decisions provide more grist and focus for the executive team to employ when it generates its opening work in the Hoshin process.

So, a strategy may form from a framework—a set of principles, assumptions, concepts, and practices that provide a way of understanding reality. Or it may emerge from a strong Hoshin review process that generates new strategic thinking. Or a visionary and entrepreneurial CEO may see the best pathway through a market, and share that strategy with the Hoshin planners via executive order. There may be other ways market positions are chosen, but this article touches on these three.

Frameworks

A CEO can use a framework to engage the leadership team in thinking the strategy through thoroughly. Models can expand and explain the reasoning behind the choice of a market strategy. And models and frameworks support the making of specific decisions around target markets, customers, price … the combination of which equal a strategy. This section touches on three such models/frameworks: industry forces analysis, value propositions and value chains, and an organization's strategic driving force.

Industry Forces

Leadership teams must understand, with some precision, the functioning of the industry in which their organization's Hoshin plan will have to succeed. A thoughtful description of the features of an industry forces a dialogue among executives on specific ways it can differentiate itself. Michael Porter's work offers questions leaders should consider when describing the industry in which they're choosing a strategy:[1]

1. How many competitors are there in the industry and how hard are they fighting with each other?
2. How much bargaining power do suppliers and customers have?

3. How how hard is it to introduce a substitute product or service in the industry?
4. How hard is it for newcomers to enter the industry?

The answers to the questions form a profile of the industry that is somewhat constant, even as players and products come and go. So leadership's work to understand the industry can help choose a winning strategy, and then that understanding can be "reused" to support decisions when adjustments to the strategy have to be made. Executive teams provide essential guidance to the organization by isolating the industry they'll work in, describing how things really work in that space, and describing a specific path through that industry where the company can win.

Value Propositions

Value propositions describe the unique things customers get when they purchase your product or service. What differentiates the product or service from any other offering? Organizations may differentiate themselves by their speed or method of manufacturing or distribution; their price, quality, or grade; or the degree of customization they offer. Tightly weaving together a set of choices can block others from profiting in the industry in the same particular way. What does your market value most, and how do you provide some of that value in ways that no one else does? Winning is about linking together choices that reduce the organization's vulnerability to competitors who are also weaving their ways through the industry.

In addition, the more limiting and integrated the choices, the more specificity is possible when Hoshin planners start their work. Leaders must be clear about what tradeoffs are going to be made to deliver the specific customer value to which they committed the organization. Clarity about the market strategy can put the organization at lower risk for producing a more ordinary or generic-seeming Hoshin plan.

Hoshin Planning as Gap Filler, Soldering Gun, and Table Saw

The value proposition is made real through a set of linked processes that get product and service to the customer and generate revenue—the value chain. If the market position sought is a new one for the organization, that chain of processes may not be in place, yet, or the current value chain may not be right. The staged map of major improvements in a three- to five-year Hoshin plan can accomplish the needed enterprise-wide renovations to the value chain. Hoshin can help fill big gaps by building capabilities that are currently missing or broken, forge stronger links among the organization's processes that deliver value to the customer, and systematically cut major areas of activity that are no longer needed.

If Porter-like frameworks were used to develop a company's market strategy, executives might launch the development of the Hoshin plan with a modified version of the traditional framing question. Something like the one below might be used to generate the ideas for the Vision Affinity.

> It is five, ten, twenty years from now. We have leveraged our relationships with suppliers, raised barriers to entrants to the industry, created a unique product/service resistant to substitutions, and provided our customers with profitable ways to leverage their buying power with us. What did we do?

The answers to this question might produce the high level categories of activity the organization will need to accomplish to build the right value chain and address the forces in the industry. Hoshin can help get an organization out in front of its competitors by disciplining the work to stage and advance system-wide breakthrough improvements that will place the company in the strategic position it has chosen. A by-product of building this narrower, planned path is the need to decline the pursuit of things that are not on it, and this can be tough. Leaders must be skilled at saying "no," despite a steady call from within and outside the organization to renege on tradeoffs that have been made and opportunities that have been thoughtfully declined.

Strategic Driver

A second angle to take when leaders work to say "no" to some strategic opportunities and focus on a differentiating market strategy is the use of a self-focused filter. Such a filter can help the organization characterize itself, making some opportunities look more obviously in its ken and others less so. Management author Michel Robert advises strategy makers to "create a clear definition of that single differentiating factor that enables the company to add unique value that competitors will have difficulty duplicating."[2] If an opportunity fits with the company's differentiating feature, it might be pursued. If it does not, it should be rejected, regardless of how inviting that opportunity looks.

The strategic drivers in Figure 3.1 are examples of organizational strategic drivers. A company might attend to more than one of these areas to some degree, but chooses one area as its defining filter. With leadership teams, sometimes the dialogue slips into a kind of wishful thinking, a debate about which driving force the organization *should* choose. If this happens, look for patterns in the actual history of decision making at the organization. What filter has been used consistently? Have competing drivers been used to accept or reject opportunities? A look back at the history of the big strategic choices made by the organization may reveal that a filter *has* been applied with some consistency, even if that filter was not clearly defined in the organization's founding or planning documents.

Embracing a strategic driver not only serves as a boundary within which a leadership team chooses a strategy, but also helps inform what the critical success metrics are for the organization and for the Hoshin. For example, if the company is primarily Product driven, then metrics like new product

Driving Force	Description
Product or Service	Strategy focused on a single product and its derivatives.
User or Customer Class	Serves only one end user or customer with a wide array of unrelated products.
Market Type or Category	Serves a common need among buyers in a market.
Technology	Rooted in a hard (chemistry, physics) or soft (expertise or know-how) technology and seeking applications of it.
Production Capability	Keep a distinctive production facility operating at maximum capacity. A 'keep it humming' strategy.
Sales or Marketing Method	Distinctive selling method.
Distribution Method	Distinctive approach to moving tangible or intangible things from one place to another.
Natural Resources	Entire purpose is the pursuit and exploitation of oil, gas, gold, timber, or other resources.
Size or Growth	Usually a conglomerate whose sole interest is growth and size for their own sake.
Return or Profit	Seeking minimum level of return or profit regardless of product or service.

Adapted from *The New Strategic Thinking* by Michel Robert

Figure 3.1 Strategic driving forces. (Data from Robert, Michel. *The New Strategic Thinking*. New York: McGraw-Hill, 2005, pp. 55–66.)

time to market and market uptake volumes might become candidates for focused tracking and process improvement efforts. Alternatively, if the company is User driven, customer loyalty and Customer Lifetime Value (CLV, in dollars) may be among the key high-level metrics. And if the company is Technology/Know-How driven, its high-level metrics might include something like technology adoption rates.

Like the work to characterize the industry, Robert's model contributes to being able to say "no" to even the most appealing opportunities. To choose a differentiating characteristic is limiting. It's a constraint, and constraints can be both a source of innovation and a way to direct an organization's resources wisely. Focused and limiting market strategies make for more focused and detailed Hoshin plans.

Hoshin Review as a Means for Spotting an Emergent Strategy

Another source of strategy can be the review information from a mature Hoshin Planning process. It's essential to conduct reviews well to understand not only what works and what doesn't in the plan, but also *why* something doesn't work well and how it can be changed or improved. Review meetings are places to share, for example, what's been learned about suppliers during an exploration for how to build a new product or service. A well-conducted Hoshin review can reveal new discoveries around how a product or service is being built, or changes needed in the value chain and the costs associated with those changes. For this to work well, there must be total transparency in the discussions, with a readiness to discuss any activities that haven't worked and to try something else.

When Hoshin strategy teams are given room to explore, they can come up with some surprises. Those surprises can be captured at review meetings, too. Discoveries made during Hoshin teamwork can be used to make adjustments to the market strategy.

Visionary CEOs and Open Style Organizations

Serendipity, experimentation, and the recognition of patterns in customer and competitor behaviors are among the variables employed by visionary leaders who create successful strategies. Their work seems, and might be at times, unconscious or instinctive, not a product of any planning methodology. Those leaders have the ability to see a profitable pathway through the marketplace when others cannot.

In addition, some CEOs create cultural environments that are freer thinking and more fluid than most. The fluidity can be further cultivated with a flexible organizational structure. Companies like Google and GitHub are known for being more innovative in organizational design, for example, allowing people to self-assign to projects that interest them or contribute to projects on which they're not formally working. The 1970s business professor Karl Weick described environments of "loosely coupled" things that produce the conditions where strategy can emerge. He used the metaphor of sand being a better sensor of wind than rocks. As a loosely coupled system, the sand can detect changes in the environment better than the larger, heavier rock. Likewise, organizations with frequent and open dialogue around strategy, and fewer restrictions on where someone can contribute ideas, may be better than too-tightly-run organizations at recognizing and responding to changes in the market. Weick and Kathleen Sutcliffe, in their book *Managing the Unexpected* offer insightful discussion and an audit tool for considering how tightly coupled and "interactively complex" your organization might be.[3]

Finally, and in a similar way, a multiyear cycle of Hoshin Planning can also contribute to the development and adjustment of market strategy. The superstructure that is temporarily laid over the existing organizational structure as part of a Hoshin Planning cycle typically runs with cross-functional teams and includes people at all levels of the organizational hierarchy. This allows for unusual movement between areas of the business, working around silos, and using an executive-mandated freedom to explore beyond usual departmental and functional boundaries. Opportunities for changes in business strategy may be spotted better in open and agile environments that value and manage movement and change.

A Test before Going "Live" with a Strategy

Leadership teams may want to evaluate their strategy before entering the initial visioning process of Hoshin Planning. Joan Magretta offers one good guide for doing this in her book *Understanding Michael Porter*.[4]

Building the Ramp Up to the Hoshin

Once the strategy for the organization is determined, Hoshin Planning work begins on an organized list of preplanning activities. The purpose of this early work is to build shared meaning and relevance from what is learned about the industry and the organization, gain eventual agreement on what the organization's priority work is, and produce a clear explanation and logic that can be shared with the entire organization as work on the plan expands to those outside the executive group.

The list in Table 3.1 resembles one that might precede the application of any strategic planning method. What makes it Hoshin-like isn't necessarily what's on the list, but rather *how* the items on the list are discussed and completed. There is an expectation in Hoshin Planning that a leadership team will reduce and prioritize as part of the development of these pieces of planning, and that when resources are limited, the Hoshin will be top priority. The skills to evaluate and reduce options, and to be sufficient but spare

Table 3.1 Preplanning List

Output	*Questions*
Mission statement	For what purpose does the organization exist? Whose need(s) are we meeting? What do we make or do to deliver value? Where do we provide our product or services? How do we deliver them?
Values	What is important to us as we employ our mission? In what kinds of behaviors do we believe? What are we each personally willing to do differently as an executive to make our values transparent to people?
Performance metrics	How will we know if we have succeeded in our mission? What will it look like numerically? What is the objective evidence that we are performing well?
Assumptions about the future	What do we think will happen in the next five to twenty years in our regulatory, technological, economic, political, and physical environment that will shape our decisions about strategy?
Customers and their demands	Who are our best customers today? Who do we want our customers to be in one to five years? How do our desired customers behave, and what do they want? What will we do to find out?
Value chain	What are the five to seven core processes that deliver value to our customers and generate revenue? What major functions does the business perform in building and providing a product or service to our customers? How well are we executing those processes now? Which processes might be phased out as we move into the future? Where are the gaps?
Environmental information	What are the regulatory changes likely in our industry? What are the demographic changes likely? How will technology changes likely impact our customers and our organization?

in the activities chosen to gather the data and information, are consciously applied. Like a good market strategy, Hoshin Planning prework seeks to do only what it must to accomplish its objective.

These items should be revisited at the start of each Hoshin Planning cycle, but they don't necessarily have to be done all over again. The items might be revisited more thoroughly, if something dramatic has changed in the organization's market strategy or the conditions in which leaders are building the Hoshin plan.

Mission and Values

Simply stated, for what purpose does the organization exist? At the start of the Hoshin Planning process, leaders need to remind people what they are coming to work for every day. A clear mission helps build the boundaries within which the Hoshin Planning work will be developed.

On values, clearly stated ones help Hoshin teams make important choices in their work. Which behaviors or practices are essential to represent in the work, and which are unacceptable? What priority values absolutely must be built in to the solutions and activities Hoshin teams produce? Some organizations wear their values in more obvious ways than others. The CEO of Aetna invites employees to practice meditation during work hours, as he does himself, and the company is one of few insurers to cover holistic health interventions, such as acupuncture and massage. In 2012 Aetna cited a 7.3 percent drop in its own employee health care costs, about $9 million in savings, and they remain about 3 percent lower than they were before yoga and meditation were introduced at the company.[5]

Milliken and Company, one of Ethisphere's 2014 most ethical companies in the world,[6] has had a decades long focused commitment to innovation. Formed and nurtured by founder Roger Milliken, the company's spirit of innovation has produced a better understanding of plastics clarification than its rivals and has shaped the industries it serves.[7] Producing plastics with polypropylene, for example, pushes others to use this lower-environmental-impact plastic. Milliken uses this technical expertise, and decades of process improvement and planning expertise, to help its business customers manufacture more efficiently, resulting in those businesses making smaller carbon footprints.

To advance these values as well as they do, Milliken's way of designing and delivering products to its customers has to be tailored in a way that would be difficult for a competitor to duplicate. Hoshin Planning, over successive annual cycles of application, can systematically build this kind of custom set of core design and production processes. David Moody, president of Milliken Chemical, says, "We use Hoshin to implement our plans with discipline and to ensure that our values are embedded not only in the strategy and goals of the plan, but in how the plans are implemented."[8] A clear and consistent expression of a company's values in the early stages of the Hoshin Planning cycle helps ensure at least two things: first, that those values inform the strategic choices the company makes and second, that as the plans at the executive level cascade down into the organization, the choices next level managers make to implement the Hoshin consistently reflect the values that the company's leadership has defined as important.

Performance Metrics

Executives must be absolutely clear about what fuels the organization's engine. High-level financial metrics are only the final measurable outcome in a cycle that begins with the face-to-face work people do with customers and the daily tasks they do that impact the customer. At the highest level of the organization executives need to define the handful of metrics that open a window for them into the performance of a critical few processes that add customer value. It's leadership's responsibility to ensure that the Hoshin plan is aimed at what is important for the organization's success. People pay attention to what gets measured.

Leaders need to define a balanced set of metrics before leaving the preplanning phase and choosing a Hoshin. Quality, cost, service delivery, and culture/morale are the four classic areas of metrics often used in Hoshin Planning, and a leadership team needs perfect consensus on what the mix of the highest level metrics of success are. Performance discussions at the executive level are essential, and the leadership of Hoshin Planning is compromised if metrics are vague, or there is not absolute agreement. All leaders must agree on a clear and objective definition of success.

Assumptions about the Future

One of the more energizing dialogues in the Hoshin preplanning work can be around the assumptions about the future. Good CEOs let the daffy-sounding predictions of the world twenty and even fifty years from now get into the mix, too. A mistaken belief that there isn't time for speculating about the distant future can be set aside to exercise the leadership team's ability to think forward, far forward. Rational vetting, a "what is really likely to happen" filter, can be applied after a group has an opportunity for an expansive dialogue about the future.

Figure 3.2 shows the assumptions NV Energy used to help create a context for their Hoshin plan. The list was completed as the company's power generation division began its first Hoshin Planning cycle in 2009. It was modified in 2010, between its first and second Hoshin Planning cycles.

Assumptions about the future
five to ten years from now

1 Intertia will build, and two utilities will merge into a single control area.
2 Gas and coal prices will be predictable over the long term.
3 Gas sources will be the same.
4 Ruby pipeline will be built.
5 PRB (powder river basin) will be coal fuel of choice.
6 Nevada will remain regulated, but renewable development projects will compete for sites, resources, and customers.
7 There will be zero to low load growth in the two utilities forcing management to change from growth to optimization.
8 Both utilities will manage O and M budget growth to "at or below" our load growth.
9 Our advantage will be that we have existing plants and plant sites.
10 Customers will have the choice of green power.
11 We will have BACT SO_2, No_x, and mercury at coal plants, and BART at gas steam plants.
12 Carbon sequestration will not become commercially viable.
13 We will not build a coal plant—may buy one.
14 Major maintenance will become condition based.
15 Headcount will not increase—FTE per mwh will decrease.
16 ~40% of generation employees will retire.
17 Workforce skill sets and utilization will change dramatically.
18 Additional generation will be built or bought in Nevada.
19 Carbon resolved for utilities—greater focus on efficiency.
20 Generation fleet will be part of the resolution of intermittency.
21 Some generating assets will be aged and retired.
22 Pressure on rates will be intense—increased regulatory scrutiny.
23 Customers will expect more direct interface with their utility.
24 We are long off-peak, off-season leading to increased off-system sales and change in operating philosophy.

August 19, 2010 modification

Original – March, 2009 Hoshin Planning – NVE generation

Figure 3.2 NV Energy assumptions about the future. (Reprinted with permission.)

Slight changes in the assumptions prompted adjustments in both the metrics around the first Hoshin and the selection of the second Hoshin. Significantly, although these assumptions were adjusted, the main elements of the five-year vision did not change, in part a credit to the quality of their preplanning work in the beginning.

Discussion of Current and Future Customers

Hoshin leaders ask their organization to reach out to current key customers expected to still be part of the customer mix in three to five years and to potential new customers, and in both cases, use an organized interviewing process to find out what those customers really need and want. As an example, Figure 3.3 outlines a nursing school's list of customers and their demands, developed at the start of their Hoshin Planning cycle. A simple matrix like this can be a good tool for leadership teams to use to start the dialogue.

Leadership teams use the customer list to talk about which customers they will continue to serve and which may need to be weaned away as part of a revised market strategy. Discussion should include numbers around how each customer type contributes to revenue and what opportunities there may be to deliver additional value to customers currently served.

Students	Responsive faculty, quality curriculum and teaching, convenience/access, $, job placement, support passing exams (certificates)
Faculty/Staff	Policies and procedures, mentorship for career development, concrete help and administrative support, want communication, want to feel valued
Employers	Workforce planning, excellent entry level nurses, reliable ready to work, socialized to position, skill set, potential for leadership, steady quality supply
Alumni	Programs they are proud of, connections
Legislators	Information, education, "go to" people, know what the issues are
Funding Sources	Innovative projects with important implications
Administration	Information/data, success, good PR, no negative surprises, high competence, well run ship, following policies/procedures
The profession of nursing	Knowledge, innovation, knowing what the issues are, bodies, volunteer leaders
Patients	Individualized expert care, cutting edge innovations, "care"

Figure 3.3 Customers and needs.

Customer Satisfaction and the Kano Model

The customer discussion ought to work toward a deep and data-based understanding of the prioritized customers' needs and interests. One way to frame a dialogue on customer expectations is through the lens of a model like that of Noriaki Kano. His work introduces the idea of "attractive" versus "must-be" quality. In Kano's model, the attributes and features of your products and services fall into one of three main categories.[9] The first category is referred to as "must-be," now usually called "expected" quality. The customer may not state these requirements, as it's expected you already know what they are and will provide them, and you'll receive no special thanks. These are considered "table stakes," minimum entry requirements for being in this business. For a Western hotel, such things as a lock on the door and clean sheets might sit in the "expected" quality category.

In the second category, "one-dimensional" quality, are things your customer asks for and also notices when you provide them. Again for our hotel patron, this might be something like a non-down pillow or a late check-in time, requested when the reservation was made. Both "expected" and "one-dimensional" quality characteristics must be fulfilled by the combined work of your value chain if you want to win or keep that customer.

In the third category are product or service characteristics likely to "attract and excite" the customer. These things are considered "exciting" quality. The customer does not expect these things. In fact, he or she may not even know to ask for them. At the hotel, that might be something like a free room night or a complimentary spa pass. These are considered "competitive stakes," things that make the company stand out relative to others.

For targeted customers in the longer view planning of Hoshin Kanri, at least three messages from Kano are important. First, the items in these categories move, often from "exciting" to "one-dimensional" to "expected." For example, maybe a decade ago, the option to slip out of the check-in line and use a lineless kiosk to get to your room was exciting. Now it might be expected as a phone app that allows us to check in from the taxi on the way from the airport. If you're not the hotel that provided this convenience first, you have to catch up now anyway, because the items in Kano's categories can sweep industries fast. You may not have enjoyed the benefit of exciting your customer first, but once a product or service feature is out there, it's likely that every player in the space will be asked for it by its customers, putting the feature in the "one-dimensional" quality category, and then at some point it will be "expected," or "must-be."

Second, leaders need to consider what risks and costs are associated with giving their product or service "exciting" quality attributes and they must also translate the goodwill earned by the move into a quantifiable expected return. Finally, leaders need to ask: What should be built into the Hoshin plan to accommodate for innovations becoming "expected"? Will providing "exciting" quality require a change in the value chain that can be accomplished in the Hoshin plan? Quality management expert and one of Kano's translators, Glenn Mazur, offers some charming translation notes (shown in callout box below) about the "exciting" quality category.

"Exciting" Quality Translation

In Kano's original article, he uses the Japanese word, *mi ryoku teki*, which he translates as "attractive." *Mi* is the combination of the character for devil (ghost) and tallest branch of a tree. In modern Japanese, it means to bewitch, enchant, fascinate. *Ryoku* is the character for sinew, strength, effort. *Teki* makes it an adjective to modify the following phrase, "quality elements." Together they mean the power to ... charm, allure, captivate, attract, appeal, magnetic, fetching.[10]

Core Processes/Value Chain

There are only a few high-level processes organizations have that generate direct customer satisfaction and generate revenue. There are also strategic and supporting processes, but here we want to know what processes are essential to getting the organization's product or service to market consistent with a distinctive value being offered to customers. This is the value chain. How these processes work must be understood to know if new processes might need to be added to achieve the Hoshin objectives or if current pieces of the chain might be made obsolete once the Hoshin is achieved. Figure 3.4 shows what a high-level set of value-adding processes might look like for a pharmaceuticals organization.

In addition, a more detailed picture of how the business runs can add critical insight into how the organization currently functions. Figure 3.5 shows an example of a high-level flowchart for a firm that is split between two independently operated divisions, one that provides analysis of patient

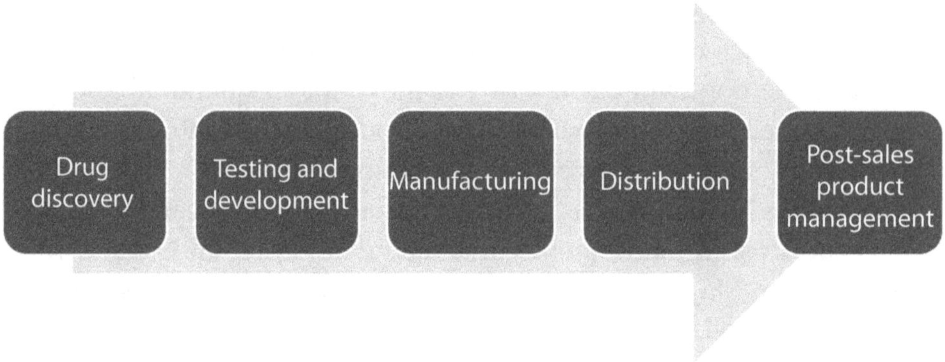

Figure 3.4 Overall business flow.

health data and another that uses that data to provide customized health coaching to reduce the effects of chronic illnesses such as diabetes and congestive heart failure.

This chart is especially elegant, with its two divisions of the organization having their own paths. What they deliver to each other is made clear through the use of pictures rather than too many words. Messy patient data (sick and well faces mixed up) are provided by clients and goes in on the right side; organized data (sick and well faces clustered into groups) are generated by Division A and delivered to Division B's database, which allows Division B to deliver contract health counseling services to its clients. Notice that nowhere are processes related to Human Resources, Finance, Facilities, Information Technology, or other support services shown. This image shows only those steps that provide direct value to customers (insurers and care providers) and to which revenue can be directly attributed.

Calling on customers, making products, and delivering the service are examples of processes that directly impact the value delivered to customers. We can differentiate ourselves by how well and at what cost we do these things. That's why an explanation of the organization's value chain and its costs should be part of the context the leadership team has in the room when it generates the elements of the vision that open the Hoshin Planning process. We might do what our competitors do to deliver products and services, but we work to do them in different and better ways. We can use Hoshin to push us to what Porter called the "Productivity Frontier," the place where raw materials, skills, and technology are applied at their highest best level.

Everyone needs to understand how the organization currently runs to appreciate how it will be the same or different after the Hoshin is implemented and to make changes that will help ensure success on the business metrics.

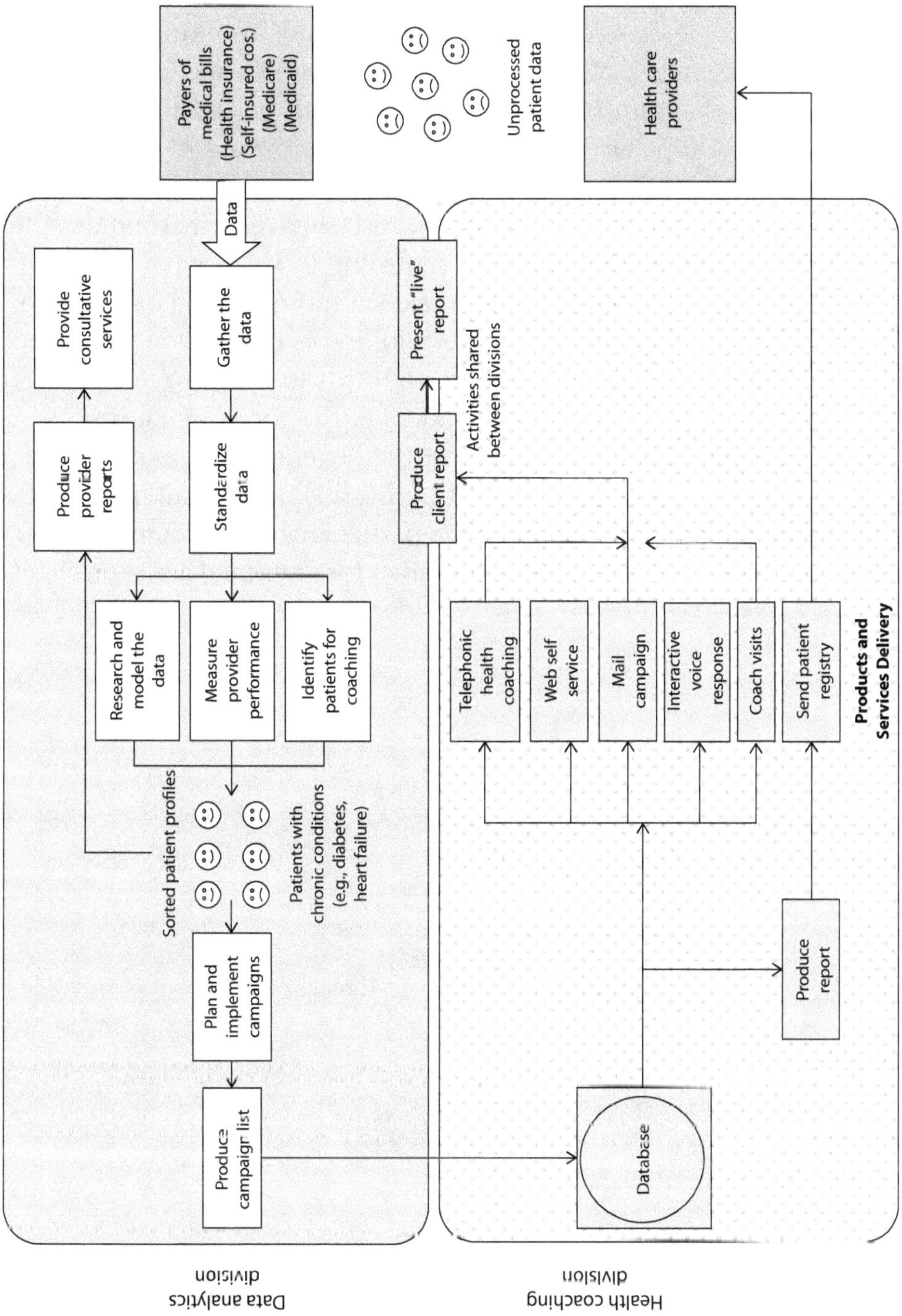

Figure 3.5 Core process flow example.

Strengths, Weaknesses, Opportunities, Threats

It's smart to discuss what the organization has to work with as it moves into new territory. One classic discussion tool for this thinking is the SWOT analysis—Strengths, Weaknesses, Opportunities, and Threats. Figure 3.6 shows a diagram of how sticky notes can be used to populate a couple of sheets of flipchart paper with an overall picture of an executive team's discussion. Figure 3.7 shows a typed example from a medical system.

SWOT seems to fall in and out of vogue. In fast-paced environments, it can be seen as a planning dinosaur, not returning in value what it costs in time to do. However, a leadership team makes a more viable Hoshin plan when some of this kind of self-evaluative thinking is done. One way to complete the important thinking that goes on with this tool is through a quick brainstorm that ensures input from all leaders in the room. What results is an informal, but credible self-assessment that leverages the expertise of the people in the organization who must be most skilled at talking about the reasons the organization has chosen a particular Hoshin. Actions to strengthen weaknesses and leverage strengths, for example, are more likely

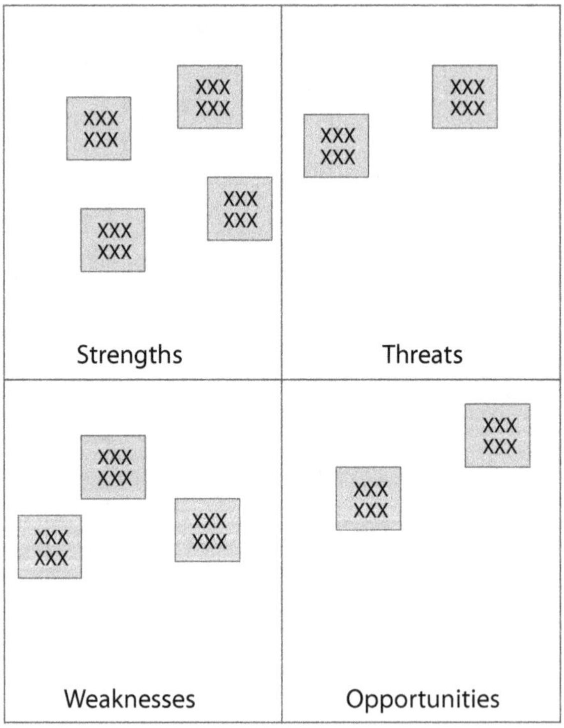

Figure 3.6 SWOT format.

SWOT
Strengths/Weaknesses/Opportunities/Threats

Strengths

- Institutional knowledge
- Proven track record of successful projects
- Leadership support for reengineered planning processes
- Enthusiasm of team
- Relationships with community providers
- Creativity
- Motivation
- Expertise and professionalism of staff
- Sophisticated data tools
- Responsive to customers
- Range of background experience to draw from
- In-house agency model already in place
- Breadth of skill sets among current staff (at this table)

Weaknesses

- Poor coordination with finance
- Insufficient MD/$ resources
- Lack of consistent process
- Take too much of the process on our shoulders
- Lack of alignment
- Lack of consistent faculty involvement
- There's never been a system in place for enforcement (have to teach old dogs new tricks)
- Pent-up demand for support
- Insufficient numbers of staff to meet needs
- Gaps in market-planning continuum
- Lack of centralized process for planning and marketing
- Complex organization
- New model/responsibilities and existing workload

Threats

- New leadership
- Competing processes and resources
- Customers will go elsewhere due to lack of resources
- Other providers in the country
- Reject radical element
- "Been there done that" skepticism re: "new" planning model
- Current culture of "run around" the planning process
- Organization gets distracted by another patient safety event
- Clients can go outside to fund their bad ideas
- Leadership cannot be disciplined
- High initial expectations (overzealous)
- External market forces create $ barriers
- Fear of change
- Turf issues related to data
- Lack of buy-in with market data

Opportunities

- New leadership—clean slate
- New decision support tools
- New info tech leadership
- A backbone/support system that will reinforce marketing communications
- Acknowledgment that the current system isn't working
- Ability to differentiate with our ability to integrate and execute ideas
- New, effective model for planning and running enterprise
- To create aggressive, nimble, customer friendly clinical services
- Weakened university
- Ability to attract additional business development experts
- Create something that propels us to the next level

Figure 3.7 SWOT example.

to be included in more detailed levels of the Hoshin plan, if they have been discussed and communicated at the leadership level.

To emphasize, this is a way for leaders to tap their own knowledge about the organization. Slowing SWOT down with research and an orientation toward data-driven statements is another type of work, which may add value, but that is not the same as leaders coming together to build a shared assessment. In this early Hoshin dialogue, leaders draw on their own observations and knowledge to develop shared insights about the organization.

Environmental Information

Organizations today often have internal groups who provide market and environmental information and analyses. These groups should be tapped early in the Hoshin Planning cycle to provide relevant contextual information. For example, leaders should have up-to-date information about what is happening in the regulatory environment and what the demographics of their target market(s) look like. Using maps and pictures to display this information can help leaders and others visualize the playing field and participate in the discussion.

In addition, and consistent with the earlier section related to "Assumptions about the Future," leaders ought to consider with some specificity what some likely changes to the organization's environment may be down the road. Some leadership teams use the STEEP (Social, Technological, Economic, Environmental, and Political) factors initially offered as PEST (without the Environmental) by Francis Joseph Aquilar[11] in the 1960s and adapted by others since then. Each area is discussed to anticipate what are likely changes in that area, and chosen responses to those changes are built into the plan. For example, leaders may prepare managers with necessary exposure and training for anticipated changes. This is a form of focused contingency planning that can make the Hoshin plan less vulnerable to being ambushed by changes in the environment.

Setting a Solid Foundation for Choosing and Implementing the Hoshin

All of these preplanning activities serve as preparation for when a leadership team creates its first Affinity Diagram on the elements of the vision. With a sound market strategy and good preplanning, a leadership team can then bring it all together in a detailed vision of the future. It's that vision that serves as the foundation for the Hoshin plan. If the organization wants to create a sustainable focus and align everyone around it, it must choose a single top priority from that vision on which to focus for a specific period of time. Finding and deploying that focus is the role of Hoshin Planning.

Preplanning work gets leaders conversant in the structure of their industry, the organization's skills and weaknesses, and the customers to which the organization will provide value. Due diligence prior to development of the elements of a future vision also establishes a better quality consensus around how success will be measured and provides an opportunity for the CEO to remind all leaders of the purpose and values the organization serves. This work can't be skipped without a cost to the quality of the Hoshin plan.

Endnotes

1. Porter, Michael E. "The Five Competitive Forces That Shape Strategy," *Harvard Business Review*, January 2008, pp. 78–93.
2. Robert, Michel. *The New Strategic Thinking*. New York: McGraw-Hill Publishing, 2005, p. 55.
3. Weick, Karl E., and Sutcliffe, Kathleen M. *Managing the Unexpected: Sustained Performance in a Complex World*. Second Edition. Hoboken, NJ: Jossey-Bass, 2007, p. 91.
4. Magretta, Joan. *Understanding Michael Porter: The Essential Guide to Competition and Strategy*. Cambridge, MA: Harvard Business Review Press, December 6, 2011, p. 160.
5. Gelles, David. "At Aetna, a CEO's Management by Mantra." *New York Times*, February 2015.
6. 2014 World's Most Ethical Companies, http://www.ethisphere.com.
7. Bussey, John. "The Anti-Kodak: How a US Firm Innovates and Thrives." *The Wall Street Journal*, January 6, 2012.
8. Moody, David. Milliken and Company, July 10, 2015, personal correspondence.
9. Kano, Noriaki. "Attractive Quality and Must-Be Quality" 12th Annual Meeting (1982) of the Japan Society of Quality Control (JSQC), p. B6.
10. Mazur, Glenn H., and Zultner, Richard E. "Transactions from the Eighteenth Symposium on Quality Function Deployment," *The Kano Model: Recent Developments*. Austin, TX: QFD Institute, 2006, pp. 109–116.
11. Aguilar, Francis J. *Scanning the Business Environment*. New York: Macmillan, 1967.

Article 4

Measuring to Support Hoshin

Contents

> A big problem in training and in leadership in the United States arises from a flexible standard of what is acceptable work and what is not.[1]
>
> **W. Edwards Deming**
> *Out of the Crisis*

A fascination with measuring things about ourselves is part of the zeitgeist of our time. There is a device you can fasten to a golf club to track and analyze your swing. There is a phone app you can use to record every time you're happy or sad and then study the resulting charts to see if you can correlate your peaks and valleys with specific things that happen in your day. We monitor our health by the day or even hour with the help of wearable technology, heart rate arm bands, watches that monitor the patterns of movement of an elderly loved one and report when the pattern is broken, a baby bootie with a pulse oximeter that measures oxygen levels, and even electrochemical tattoos that will provide you running data on your body's chemistry. Lifelogging, described as "self-knowledge through measurement," is now part of the modern daily life. Author Bruce Feiler summarizes the "quantified self" movement.

> Words take a back seat to spreadsheets and metrics as more aspects of life become quantifiable and apps track even our moods. It looks like the nerds have won. The hovering eye in the sky watching every move you make: it's you.[2]

Despite the general mood of monitoring, though, the focused, skillful, and broad use of measure to monitor an organization's progress on goals can be elusive. The quality with which leaders identify the right metrics, gather data, and teach others about measurement varies, and being good at measuring is central to completing a high-quality application of Hoshin Planning. Best practitioners of Hoshin learn how to interact with data. They develop a facility for recognizing in numbers an indicator of whether what's being done day-to-day is moving the organization in a desired direction or not.

Having sprouted from the rich measurement environment of statistical process control, the early manifestations of Hoshin Planning featured measurement and analysis of critical business results prominently at the opening of the planning cycle. Hoshin Planning pioneer Yoji Akao wrote, "To set annual objectives correctly, companies must review the previous year's performance. The appropriateness of targets and means set this year depends on how thoroughly the results of last year are reviewed."[3] Akao, with other early Hoshin planners, emphasized the importance of objective performance analysis in defining Hoshin plan priorities. Figure 4.1 shows check marks where leaders might pay particular attention to measurement concerns in the Hoshin Planning cycle.

Perhaps today the pace at which a company's market can change makes relying exclusively on last year's performance information an insufficient way to select a Hoshin. The possibility of game changing technology, for example, makes an industry-wide revolution in how its players do business more likely. Still, innovation and exciting market maneuvers alone, without a corresponding focus on measurable operational excellence, may produce disappointing business results. Using historical data on the performance of processes core to the business, although not entirely sufficient, is still critical to the selection of the Hoshin.

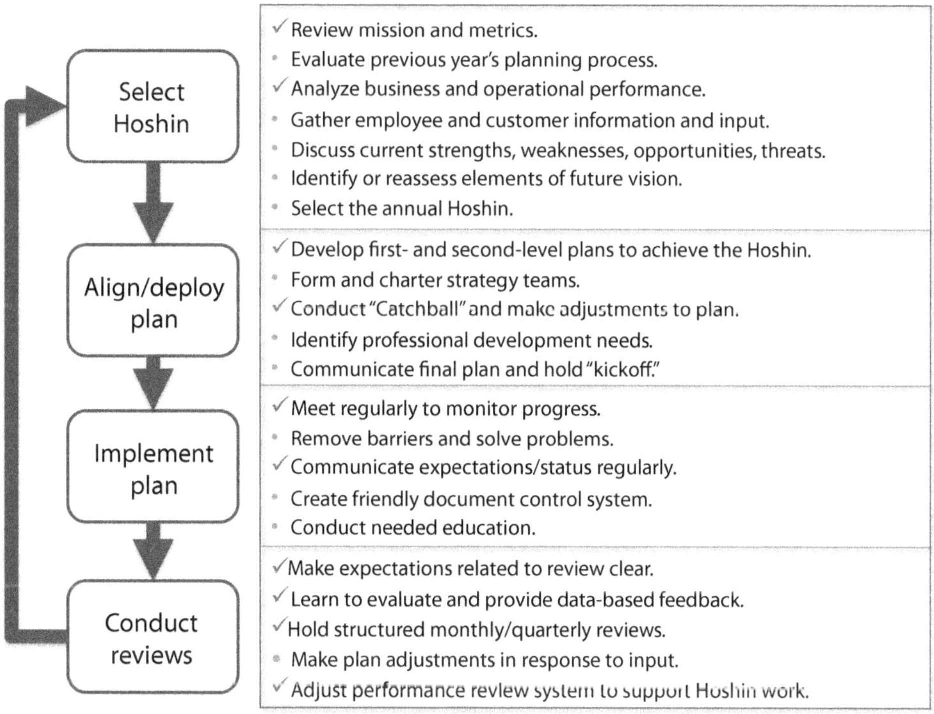

Figure 4.1 Hoshin Planning phases and actions—measurement.

Metrics as a Pathway to Consensus about the Business

In Hoshin Planning, the purpose of measurement is at least threefold. We measure to agree on an objective definition of a desired end state; build understanding around an area that is new to the organization—a breakthrough; and ensure that our strategic choices and related activities are meeting the organization's business objectives.

Leadership Consensus through Metrics

- Building a shared objective definition of a desired end state.
- Learning how to measure what the organization hasn't done before.
- Integrating Hoshin metrics with business fundamentals.

Building a Shared Definition of Success

The evidence of a consensus decision is that once it is made, everyone who participated in making it begins to behave consistent with the spirit of that decision. The results of a Hoshin plan are influenced in part by whether consensus is achieved along the course of its selection and implementation. One means for arriving at consensus is the dialogue that takes place to create an objective definition of success.

"We had to be more clear about exactly what we meant by an outcome for our Hoshin. I was surprised that when we thought we meant the same thing, it turned out we meant different things, and it showed when we tried to put a metric on it,"[4] stated Julia O'Neill, Hoshin program manager for Merck Vaccine and Production Technical Operations (VAPTO). The VAPTO Hoshin was to *Build an Integrated Technical Network to Provide "One-Stop-Shop" Solutions.* Initially, the leadership team generated the brainstormed list of metrics shown in Figure 4.2. In a second pass, the team refined the list, but they still disagreed about what was meant by "network." To some, it referred to a linked set of experts and services the VAPTO team could bring to bear for its clients. To others on the team, it was more of an online repository for names and numbers, set up in an org chart.

Through a third dialogue, the leadership team reduced the list of metrics and was focused on ultimate outcomes. They agreed that the Hoshin was about building networking capability in service to project excellence, a more complex contribution of value than a mere library of resources. The final list the team used for its high-level annual plan table is shown in Figure 4.3.

Hoshin metrics (draft 1)

Provide integrated "one-stop-shop" solutions

- ✓ Installation
  - ✓ Participation in external organizations (membership, committees)
  - ✓ No. of publications and presentations by team members
  - ✓ % completed—map of global technical network
  - ✓ No. of connections made between projects in our group
- ✓ Realization
  - ✓ Successes in building our external networks (how do we measure?)
  - ✓ Efficiency: person-hours saved, FTE savings, project turnaround time
  - ✓ No. referrals to technical integration group—% repeat customers
  - ✓ Annual voice of the customer survey results
  - ✓ % projects with A3/RACI/standard contract

Figure 4.2 Merck VAPTO first draft Hoshin metrics. (Reprinted with permission.)

Hoshin metrics (draft 2)

Provide integrated "one-stop-shop" solutions

- Execution accuracy
  - (% projects delivered on time, % variation vs. target date)
- Person-hours saved
- %/no. of referrals/returning customers to technology integration
- % projects with A3/RACI/standard milestones/contract
- Annual voice of the customer survey results
- % of network in place

Figure 4.3 Merck VAPTO later draft Hoshin metrics. (Reprinted with permission.)

Leaning Out—Learning How to Measure What the Organization Hasn't Done Before

If the Hoshin is a bona fide breakthrough, it's likely to involve the application of designs and processes that are unprecedented for the organization. As a result, the organization's ability to measure performance may be limited at first. Malcolm Isley, vice president of strategic planning at Greenville Health System (GHS) says it like this, "Something like VAP (hospital infections from ventilator associated pneumonia) is relatively easy. It's granular; you're measuring something discrete. And there is a lot of outside data to compare and learn from. But the more strategic the goal is, the harder it is to put a measure around it."[5]

John Frankovich, Hoshin program manager at NV Energy, expresses the question well, "How do you identify a simple measurement that's going to measure what we're actually impacting when we implement our *Workforce Re-design* Hoshin?"[6] They used man-hours per megawatt, and then refined it to more nuanced measures over time. Leaders can start by asking "What is the evidence of a system-wide performance issue right now?" and "What can we count or measure that will provide objective evidence that we've achieved a breakthrough level of improvement?" These questions can provide a framework for a leadership dialogue, the outcome of which can be a set of metrics for the Hoshin. Table 4.1 shows some examples of high-level Hoshin metrics.

Leaders need to lean out with metrics as they lean out with the Hoshin objective, which should be getting the organization out in front of its own current practices. As an example, Time-Based Activity-Based Costing (TBABC) is an evolving way for healthcare providers to measure the cost of care to a patient all the way along that patient's care path, even as that patient moves between the different providers in the health system's network. It's a more informative of the way care is delivered today, but getting

Table 4.1 Examples of High-Level Hoshin Metrics

Industry	*Hoshin*	*Metrics*
Power generation	Work management	No. of break-in hours
Pharmaceuticals manufacturing	Provide "one-stop-shop" services to our customers	• Person-hours saved • % Return customers • % Improvement in site-based voice of the customer (VOC) survey scores
Small manufacturing—laser fabric cutting	Design end-to-end solutions for fashion industry	No. of customers using all products in suite
Academic medical center	Integration of medical practices	% Patients receiving "shared care"

the data requires jumping the fences between care delivery organizations in an integrated health system. In this case, the existing tracking and reimbursement systems don't really work for getting this data. If your Hoshin objective is on the front of your industry's development curve, though, you have to find a way to track its performance, *even within a measurement system that cannot currently do it.* This is difficult, but important if the value of the Hoshin Planning work is to be valued properly.

Industry metrics are being renovated as the ways of making product, delivering service to customers, and leading organizations are evolving. Another example of this is the Research Quotient (RQ) metric developed by Anne Marie Knott. This is a new way of evaluating the return on a company's investment in research and development. As a Hoshin evolves the way the organization works and uses its resources, the way success is measured must evolve too.

Information Technology in Support of Hoshin Planning

Among the support for leaders adjusting their measuring systems to fit the breakthrough thinking that accompanies Hoshin Planning are information technology expertise and the skills of executives in deploying it. Table 4.2 suggests five characteristics that may help Info Tech systems move more closely with an organization's evolving measurement systems.

Very little, maybe none, of the kind of enterprise level measurement needed to support Hoshin takes place without the employment of information technology. Today Info Tech that is particularly helpful to the tracking

Table 4.2 Characteristics of Information Technology That Support Hoshin Planning

Characteristic	*Why Is It Important?*
Common metric definitions and templates across the organization	Supports the ability to compare "apples to apples"; contributes to breaking down departmental silos.
The ability to track data close to the customer	Keeps the Hoshin work grounded in customer behaviors, particularly if the Hoshin brings the organization into a less familiar position in the industry.
Inclusive of financial *and* nonfinancial data	Provides more balanced, earlier evidence of change as the Hoshin is being deployed; financial-only metrics are often at the tail end of a change cycle.
Accessible to the people on the ground generating performance data	Supports the broadening of the organization's skills in measurement; engages more people in the Hoshin; adds some reliability to the measurement system by locating the data gathering closer to where the work is being done.
Supportive of easy extraction and reporting by those involved in the Hoshin plan	The ability to access data and use it in plan review allows more independence and agency among Hoshin Planning teams.

of Hoshin metrics must be capable of reaching across the same organizational boundaries the Hoshin is reaching across. Data gathering and reporting silos ought to be brought down, common definitions and templates introduced, and the ability to track relevant metrics pushed out and down in the organization. The generally faster paced work of the Hoshin teams is then more likely to be informed in a timely way by changes in the performance data caused by their work.

Integrating Hoshin with Business Fundamentals

Hoshin Planning aficionados sometimes take some teasing for being overly fond of fancy matrices, but some way of showing how all the measureable goals and activities of the organization are related is essential. Senior leaders need a visual that shows how their business, Hoshin, and operational goals impact and support each other. Figure 4.4 shows a partially completed matrix for a nursing school that is part of a prominent academic medical center.

Correlation
⦿ Strong
O Medium
△ Some

School of Nursing

			Strategic Objectives									Daily Management Objectives												Measures	Goals	Responsibility
			Hoshin			Other						Critical Processes					Critical Success Measures									
			Model of Operational Excellence			Interdisciplinary cooperation	Clinical technology development	Area employer of choice	Contribute to global health			Faculty development and teaching	Research and Publishing	Clinical Care	Recruitment	Student body care	National rankings	% variance to budget	% income growth	# publications	% grads cert/licensed	Scores work climate survey	Scores student exit surveys			
Strategies	Hoshin Strategies	1.0 Process improve method selection	⦿																					Date		
		2.0 Train and communicate	⦿																							
		3.0 Identify and improve processes	⦿									⦿														
		4.0 Evaluate processes	⦿														△									
		5.0 Match infrastructure to new processes	⦿															⦿								
	Maintain	Teaching performance management system				O						⦿	O			O	⦿			⦿		O	⦿			
	Improve	Database mgmt & integration				O	⦿	△																		
		Student clinical placement process				O									O	⦿	⦿				⦿		⦿			
	Projects	New research building											⦿													
		Develop PhD program				⦿							⦿				⦿			⦿						
		Measures																								
		Goals																								

Figure 4.4 Strategic and operational goals.

The symbols show where annual Hoshin objectives intersect or are supported by operational improvement activities and how strong the relationship is. Displaying all the organization's main resource demanding activities together like this can help show where there are gaps in the critical processes for supporting the transformative work the Hoshin initiates, as well as what critical processes and metrics may not be well-correlated to the evolving work of the organization and might be candidates for retirement.

Types of Hoshin Metrics

What are chosen as metrics for the Hoshin plan can shape the behaviors of people in the organization. Most people in organizations pay attention to what gets measured. Fewer, more carefully selected metrics is better than having a cover-all-the-bases way of monitoring performance. Watching too many dials can be time consuming and disorienting to people. Leaders need to work toward a set of metrics that hold people accountable to a focused set of strategic goals.

There are some traditional ways of talking about metrics in the Hoshin Planning process. Table 4.3 refers to six that are touched on in this article.

Table 4.3 Some Types of Metrics and Their Uses

Type of Metric	*Use*
Nonfinancial	To move measures away from just end results to things that show up earlier in the change process
Quality, cost, delivery, morale	These are the classic Hoshin Planning measurement areas. Effort is made to create a balanced way of scoring the Hoshin's progress
Product *and* project metrics	The quality of the *way* we do things can be as important as the quality of the output; process quality metrics may provide better clues than product metrics for the sustainability of improvements
Control items	To hold gains, monitor the detailed causes of problems identified in the assessment that precedes the selection of the Hoshin objective
Interim or leading	To have early warning signs when the Hoshin, and other key high level metrics, are heading off course
Emergent metrics	To stay agile in response to changes in conditions, to keep the metrics relevant to current conditions

Beyond Financial Metrics

Financial success for an organization can produce the understandable assumption that the organization is also led well or is particularly innovative or prescient about the future or otherwise savvier than organizations with less robust financial success. Rosenzweig referred to this extending glow of success as the halo effect.[7] However, a company's financial wizardry will not necessarily reflect a core of quality work or sustainable strategic choices. Financial measures can be more readily accessible. Less transparent, not only to outsiders, but also sometimes even to those working in an organization, are the measures around processes that lead to and ensure that financial success. It's those metrics that Hoshin Planning leaders may have a harder time pinning down and tracking.

Mike Cowley, former division manager at the Hoshin Planning pioneer Hewlett-Packard, says it like this, "You have to have a way to account for the many things you have to do to make sure you get the right overall results from the Hoshin. I think a lot of the money people don't appreciate this. They lose sight, because for them there are only five numbers you have to worry about. It's anything but simple, though, when you get down to how the business is being run day-to-day."[8] Financial metrics are variables which are dependent for achievement on the management of other variables that are tracked and managed deeper in the organization's work.

Quality, Cost, Delivery, Morale

When defining the high level metrics for the Hoshin, leaders might start the dialogue with the classic daily management measurement areas as a discussion guide: Quality, Cost, Delivery, and Morale (Q, C, D, M). To start the dialogue, can the team identify a metric or two in each of the four areas that reflect success in meeting the organization's mission? These would be the highest-level metrics to which the Hoshin Planning work must contribute. Organizations sometimes grow lopsided in one area, emphasizing measures of cost, for example, and neglecting those around service. Leaders look for a balanced set of metrics at this level. Those highest-level metrics then inform the metrics of the Hoshin plan. Again at this level, using QCDM as a guide can help ensure a balanced gauge for how well the business and the Hoshin are going. Table 4.4 shows some examples of QCDM metrics.

Table 4.4 Generic QCDM Examples

Metric Area	*Examples*
Quality	Error rates—e.g., infection rates in hospitals, product defects in manufacturing Scrap and rework Customer satisfaction
Cost	Budget variance Person-hour per production unit Earnings before interest, taxes, depreciation, amortization (EBITDA)
Delivery	Turnaround times Customer complaint/inquiry response rates On-time delivery rates
Morale	Employee turnover Employee satisfaction survey scores Absenteeism Safety incidents Grievances

Product and Project Measurement

Leaders can gain important insight by measuring both the deliverables (product quality) of the Hoshin work and the planning and implementation processes that produce those deliverables (project quality). The Project Management Body of Knowledge (PMBOK) distinguishes between "Product" and "Project" metrics. Project quality measures are more important than sometimes assumed, because they can provide a good window into the sustainability, reliability, and efficiency of the new work processes the Hoshin may introduce into the organization. Project metrics are also especially important cues for improvements made at the strategic level, because they help leaders understand the organization's development stage in terms of the use of planning and improvement methodologies.

Figure 4.5 offers a set of project-related rating scales for the Hoshin Planning process. Leadership teams use these to evaluate their Hoshin work quarterly, or annually, and use the shared profile that is created to discuss specific action items for moving low scoring items higher on the scale and ensuring that high-scoring items stay high. The quality of the planning process does influence the quality outputs over time and the organization's ability to use similar planning processes in the future.

Understanding of the Core Business	Little or no understanding throughout the organization of the core business or its priorities.	1	2	3	4	5	6	7	8	9	10	Clear understanding, across the organization, of what business we are in and what the priorities are.
Breakthrough Thinking	People are stuck in old ways of thinking. New ideas are rejected without consideration.	1	2	3	4	5	6	7	8	9	10	People are flexible and open to new ideas. Innovative ways of doing things are sought.
Decision Making Practices	Decision process not clear. Decisions made by side groups, outside an agreed-on process, late, not at all, or without relevant data.	1	2	3	4	5	6	7	8	9	10	Decisions made with data, in a timely fashion, and consistent with the objectives of the Hoshin. Decisions supported by most.
Plan Review and Control	Plans not reviewed. Metrics describing success do not exist or are in conflict with each other.	1	2	3	4	5	6	7	8	9	10	Plans reviewed regularly. Metrics are documented for all strategies. Metrics are aligned vertically and across the organization.
Authority and Accountability	Confusion about who to go to for a decision. Excessive number of approvals needed. Time, deliverables, and behavior commitments not kept.	1	2	3	4	5	6	7	8	9	10	Always clear who is accountable for what. Lines of authority are set in a way that supports plan implementation. Commitments kept.
Information Sharing	Important information hoarded or over-edited before sharing. People not informed of decisions or new developments that impact their work.	1	2	3	4	5	6	7	8	9	10	People share new developments with others. People solicit, respect, and use each others' input. Leaders share decisions and information as soon as is practicable.

Figure 4.5 Rating scales for the Hoshin process.

Control Items

Metrics that are influenced by the actual performance of processes were identified as control items by early Hoshin scholars, such as Kaoru Ishikawa. Ishikawa emphasized that when those metrics go wrong, we check for results, and we also check for the cause of not achieving the results. It's that second step of cause analysis that might change the thinking around the plan and build a case for changing metrics or means. Hoshin planners take these control items as messages for how well the Hoshin plan is going and what parts of the plan might need to be adjusted.

"Voluntary control" sounds like an oxymoron, but it is the idea that if the item is set well and performance on it is made transparent to people, they will manage the quality on their own. Yoji Akao explained it this way.[9]

> The principle of control is based on the idea that people are inherently good and want to perform optimally; therefore control must be voluntary. Voluntary control must satisfy the following conditions without exception.
>
> 1. There must be a clear understanding of the target to be achieved.
> 2. There must be a clear understanding of what steps are needed to achieve the target.
> 3. There must be a clear understanding of what steps are needed to address not achieving the target.
>
> The latter two points need criteria to determine abnormality and handle anomaly.

It's a management skill to lead the Hoshin strategies toward meaningful measurable outcomes and to not simply exercise people in tracking and reporting metrics. Education and coaching work. Leadership needs to use creativity and charm at times to convince people that achievement of those particular metrics is essential to their shared success.

Mike Cowley emphasizes the criticality of hitting target metrics. "The truth is you've gotta meet the metrics, it must be done, and it's painful. I'm still a believer that, even though you can't think of a way to reach a goal on a metric sometimes, you have to do it."[10] Voluntary control does not equal the optional conformance to targets. Leaders need to get to a place where people are less dedicated to the production of a report they think their manager wants to see and more jazzed about finding creative ways to achieve bold targets that will take the organization forward.

Interim and Leading Metrics

In addition, it's a good idea to put in place metrics that provide early measureable signs that a Hoshin, or a business, may be about to run amok. These early signposts can be referred to as leading indicators. In the Hoshin plan, the metrics associated with the strategies and tactics are leading indicators for the Hoshin. Similarly, interim metrics are those that lead to or contribute to metrics that may not be expected to show results until further into the future. By watching the performance of the items on the smaller arrows, early signs can be seen and a failure at a higher level averted.

Cascading Metrics

There is a cascade effect with the metrics in Hoshin Planning, the metrics at the executive level spilling over to lower platforms in the organization, so that the pool of data at the very bottom reflects the priorities at the top. A practical way to make that spill happen from the top to the bottom is through the planning tables.

The planning tables show target metrics and specific strategies (means) for achieving those metrics. The language "targets" and "means" is used to explain the relationship between the numerical goals set (targets), and the specific activities or approaches that are identified to achieve it (means). Figure 4.6 shows one way the tables can work to break the targets and means down into more and more detail, cascading them to people working at more detailed levels of the plan.

Another way of visualizing how target metrics for an operational Hoshin improvement are cascaded is with the Flag Chart. The Flag Chart, a vintage Hoshin Planning chart with little flag images on it, will show how the control items mentioned previously are deconstructed in increasing detail across a problem area, allocating improvement targets by degrees and where appropriate. This kind of chart is also constructed on a Fishbone Diagram architecture, and punctuated by Pareto (the bars are the flags) and Run Charts (see Appendix for more on Fishbone, Pareto, and Run Charts). First used at the Japanese manufacturer Komatsu Company in the mid 1960s,[11] the Flag Chart assigns target measures among the parts of a process or production area that are contributing to the problem.

There is the saying "raised a red flag," which refers to something that stands out as a problem. Where there is a significant problem in an organization's performance data, that area is a candidate for the application of Hoshin Planning, and it's also an area where a set of "flags" might be raised. A unit's level of contribution to the problem is indicated by the size of the flag raised in its location on the chart. On the chart itself, results are emphasized, not the explanation of causes. There is typically nothing written *in* the flag.

With the military rigor of early Japanese-style Hoshin Planning, the vocabulary of "deploying a target" is used. In Figure 4.7, a manufacturer of

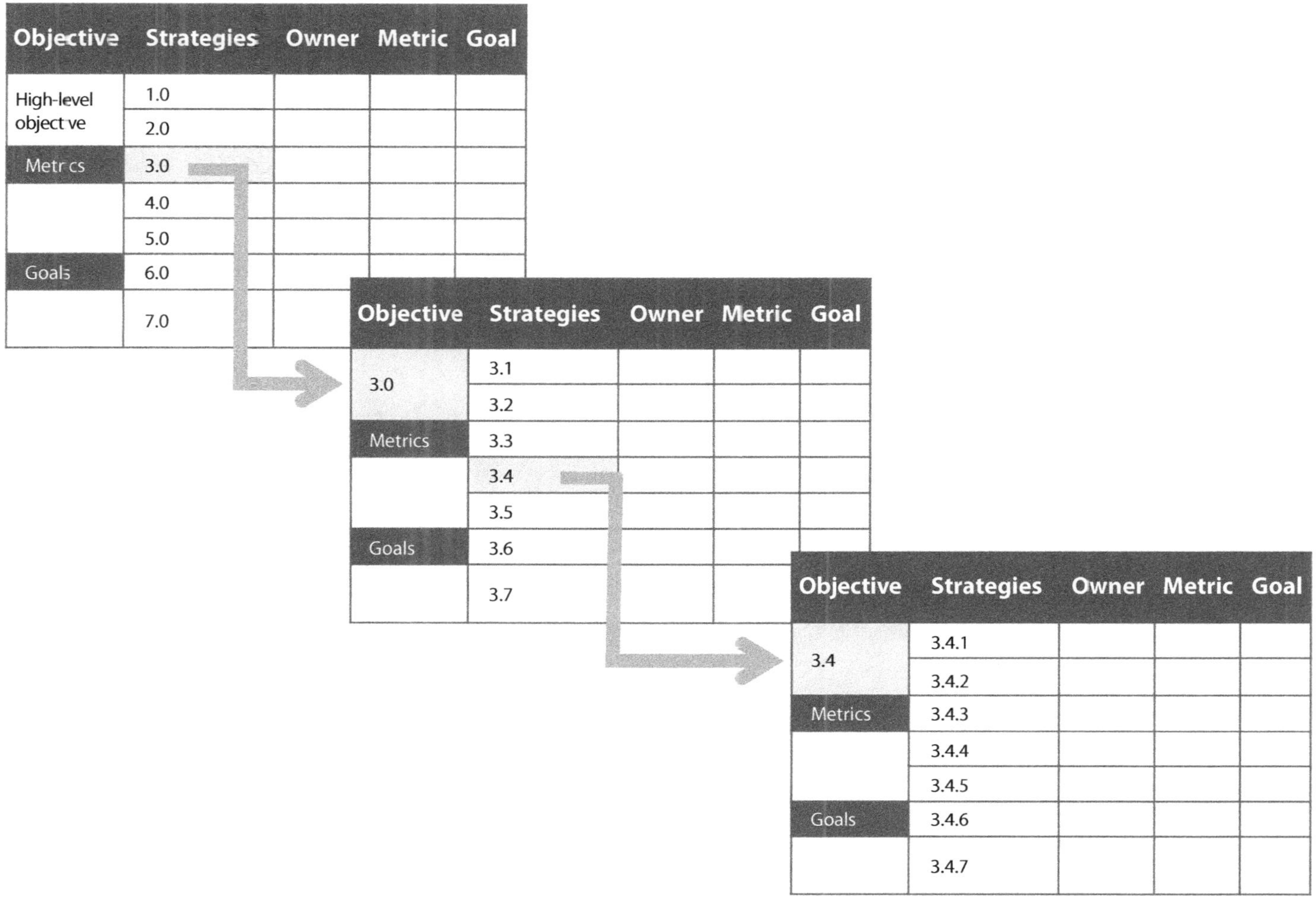

Figure 4.6 Tables flow.

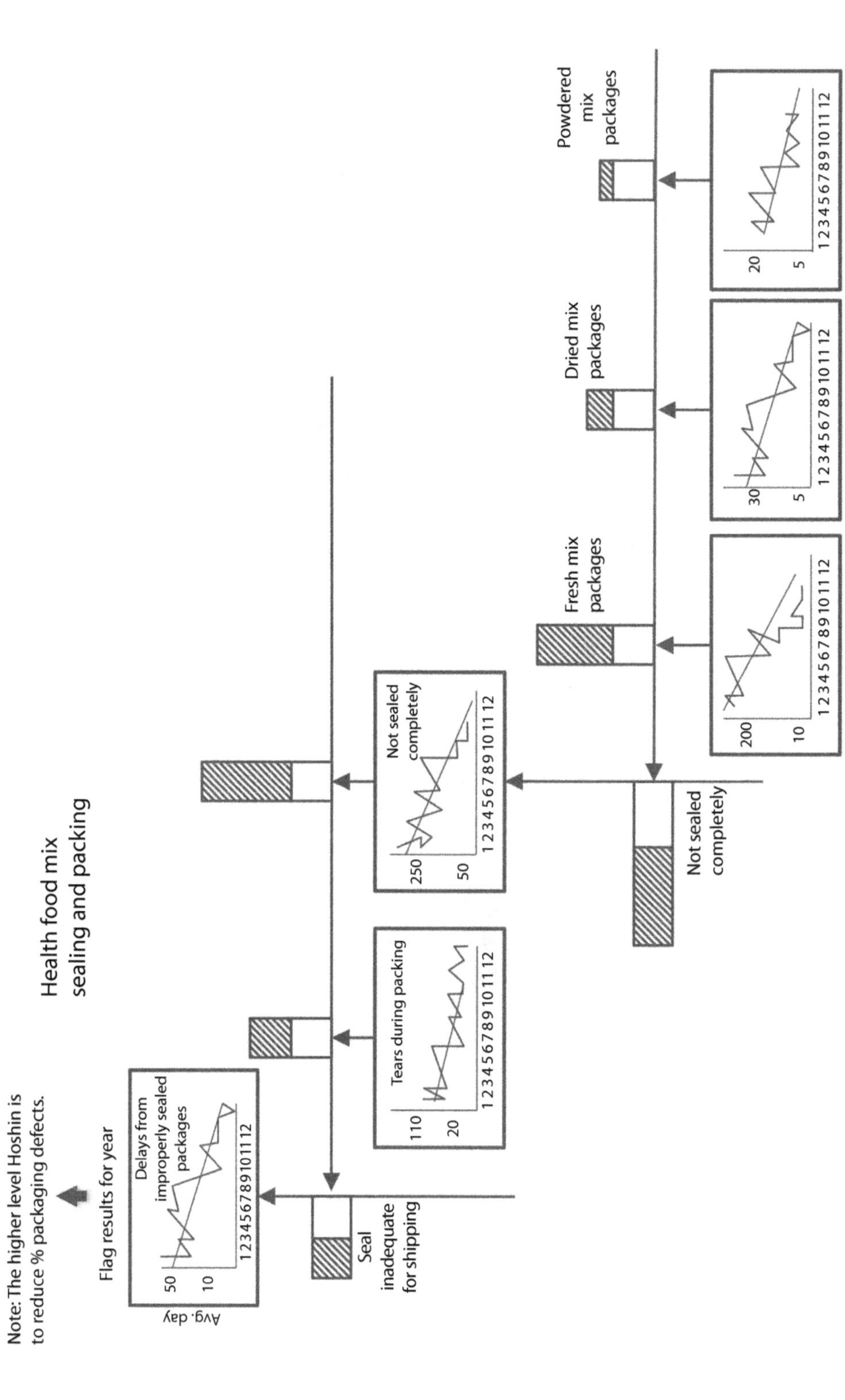

Figure 4.7 Flag Chart example.

packaged health foods has created a Flag Chart to monitor data related to packages that are not sealed properly before shipping. This is an expensive problem, involving rework and scrap, and fixing the problem could be a Hoshin-level priority. The relative sizes of the darkened areas of the flags show that packages not being sealed properly are a bigger problem than the packages tearing during packaging. In addition, the chart shows that the improvement required was greater for the packages with fresh ingredients than for packages with dried fruit of powdered protein mixes.

In this case, it doesn't make sense to assign an across-the-board improvement target of say 10% reduction of errors to the various product groups. The Powdered Mix product group should probably aim their target at zero defects, and the Fresh Mix group may need an improvement of 50% or more. The Flag Chart allows a target metric to be deployed with more specificity to where the biggest part of the problem lies.

When the improvement targets are mapped out, there are little flags on the chart showing where people will be fighting the hardest improvement battles. When those flags are all laid out, the target might be considered "deployed." When the chart is completed with the actual data shown in the small Run Charts, leadership can see where the contributions to improvement came from more specifically.

Emergent Metrics

When we choose metrics to gauge the accomplishment of a strategic breakthrough, we are trying to measurably anticipate the future. This is tricky. Again, we're doing something the organization has not done before, and we're trying to apply a measureable target to which people will be held accountable. Here is where one of the real values of the Hoshin Planning process shines. It's through review that we allow changes in what we will measure to emerge when the conditions in which the organization is planning change. "When you try to measure things that are multiyear it becomes harder, because we could change course. If you're changing over 30% of your goals, you didn't set them right. But 10–20% of our goals changing, *that* we expect,"[12] offers GHS's Isley.

Not only might the metrics evolve during the course of the Hoshin Planning cycle, but the means for achieving those metrics may need to change, as well: "You cannot necessarily achieve targets by following the implementation items planned at the beginning of the business year. You must also continue to perform factor analysis on processes and to seek proper means by which to achieve the target."[13] The planning process maps from early Hoshin practitioners like Ricoh often included an adjustment to "changes in circumstance" right in the sequence of steps. Regular review of metrics helps ensure that what is being measured is still relevant to the current situation and strategy.

Valuing Measurement Behaviors as Well as Results

Key metrics should show up on every Hoshin Planning document, as a guiding reminder of what impact the Hoshin work is intended to have. In addition to detailing the strategies and tactics that will achieve the targeted results, leaders need to mind the behaviors and practices that develop around how those results are tracked and shared. Building healthy measurement behaviors in the organization can be a valuable by-product of the Hoshin Planning process. There are some challenges to building this kind of capability, four of which are touched on here.

Challenges Using Metrics in Hoshin Planning

- Reliable data gathering
- Consistent leadership
- Measurement anxieties
- Inconsistent presentation of metrics and data

Reliable Data Gathering

As mentioned earlier in this article, reaching out with a breakthrough Hoshin objective may call for some fundamental changes in how measures are assigned and data gathered. Leaders ought to avoid a status quo approach on legacy metrics or data management systems, if those systems no longer reflect an accurate assessment of performance. It may be that the direction the Hoshin takes the organization in will require change.

It's usually not enough to mine data from the organization's processes; those data have to be processed. In addition, performance data can renew almost constantly, each repetition of a process producing another bit of information about the performance of a process. It's possible to have data you did not intentionally track turn up on its own and be very useful. Finally, it's possible to do a boring hole into the thick of key organizational processes looking for a message, and come up with nothing. Gathering and using data can be messy, and usually requires some iteration to do well, particularly at the start of a Hoshin Planning cycle, when the work being done is often new to the organization.

Data on whether a plan is on track or not are often gathered by people who may need some explanation of how to gather and report that data. Weaknesses in the organization's ability to gather data from its processes can show pretty quickly in the Hoshin cycle. "The processes that we put in place at first to collect data on the Hoshin plan were not rigorous enough to define the measurements at lower levels of the plan,"[14] stated Rich Winters, process improvement manager at NV Energy. In one case, people in the power plants accountable for reporting break-ins—number of hours of emergency or unplanned work—were dependent on others for providing job-by-job information, and the mechanism for collecting that information was at first both unclear and not seen as authoritative, something that *had* to be done.

Reporting processes must be set up and invested in up front, as part of the rest of the planning for the Hoshin. "In many projects, developing reporting tools was Day 2, and Day 2 never came. In addition, sometimes people don't have a good understanding of statistical sampling and process control, so they collect too much data or the wrong or incomplete data, which is time consuming and expensive. Those problems can all make it harder to sustain data monitoring processes and can lead to decisions based on bad or wrong data,"[15] stated Lois Gold, vice president for global project management at MetLife and former medical products' global quality manager of sales, service and administration at the Hoshin Planning pioneer Hewlett-Packard.

It is important for leaders to clearly define and put in place mechanisms to follow the measureable performance of the processes and deliverables the Hoshin process produces or changes. This work should be led by the executive team and implemented by the Hoshin program manager.

Consistent Leadership

Leaders help the organization build its measurement skills when they sustain an expectation that the organization will make data-based decisions and people will be held accountable to measureable results. "People took the transformation to a performance culture seriously, because the CEO did performance reviews with his direct reports. The CEO attended the 5th day, Friday, of every class given to managers on performance management for three years. The message got across that he was serious,"[16] says Lois Gold of MetLife. Key metrics for the Hoshin and other strategic priorities must be built into the performance management processes the organization uses to focus and support people's work. In addition, at the executive level, metrics have to be shared, a common set of quantifiable criteria to which all parts of the business will perform.

If the C-suite says one thing is important but measures and manages with something else, people will notice and will pay attention to the thing that is being measured. When choosing metrics for the Hoshin plan, leaders must assure themselves that the means for collecting and reporting performance on those metrics have been put in place and then consistently talk about those things when they meet with Hoshin work groups.

Measurement Anxieties

Exposure Phobia

Again, people pay attention to what gets measured. To have a metric attached to your area of work is to have people focused on you. If someone has a "flag" on her or his process, it can be uncomfortable. Is it any surprise, then, that we sometimes meet reluctance when we ask people to define how their work's impact can be quantified?

"I'm going to stick my neck out and say that 80% of the reason key performance indicators (KPIs) aren't forthcoming is behavioral, not a gap in technical understanding. The minute you put a number with a name and a red/green/yellow, you are going to introduce accountability and behavior changes,"[17] says John Henderson, quality director and Hoshin program manager for Gerber Scientific, Inc. Things that can help ease that sense of personal exposure include applying the same expectations to everyone and reporting business KPI's as process and department metrics rather than related to any one person.

"Too Busy" "Too Much"

There is a very real maximum threshold in an organization for how much data gathering can go on while still continuing to do what is perceived as the real work. And in a curious phenomenon of downwardly mobile expansion, a bare few metrics at the executive level have a way of becoming a large number of metrics at the more detailed levels of the Hoshin plan, which can then blossom into a massive number of metrics for mid-level managers.

And the metrics expansion phenomenon can work in the upward direction, too. People working the 1.1.1.1 level of the Hoshin plan, say, can include ten or fifteen metrics on their Implementation Tables (see the Appendix for more on Implementation Tables), two or three for each of their activities. They join other strategy and task owners at that level in rolling their metrics back up to someone who owns a higher level of the plan, say the 1.1 level. Those 1.1 level folks might receive dozens of metrics from lower levels. This can produce a leaden load of measurement that can weigh down the flow of work around the Hoshin.

One thing Hoshin Planning teams at the detailed task level can ask is "What is the one thing we must watch carefully that will tell us our work is producing a result that will move a metric dial one level up?" Try the exercise of choosing just one, but a really good one. Teams can test how good that metric is in part by asking, "If this items changes, either up or down, what will be the influence on the objective of the Hoshin?" If the answer is "none," maybe that metric can be eliminated. "Someone has to say 'no mas,'"[18] says Lois Gold. In addition, the program manager must take a look across all the Hoshin strategies and objectives and winnow down again, look for repetition among the metrics, and when possible set up centralized shared tracking to lighten the reporting load on individual strategy owners.

Inconsistent Presentation of Metrics and Data

Allowing a "Smoot"

In 1958, the living body of Oliver Smoot was used as the unit of measure for determining the length of the bridge linking the Massachusetts Institute of Technology in Cambridge with the city of Boston. The bridge is around 364 smoots long, a piece of data determined over the course of a couple of hours and with the help of his fraternity brothers.[19] Interestingly, Smoot went on to become the American National Standards Institute's chairman, a position that necessitated the consistent use of universally recognized standards, which probably didn't include the "smoot" metric. It's important to have clear operational definitions for every metric, and the more recognizable they are, the better. Recognizable units of measure, standard methods of measure ment, and some conventions for reporting data help make Hoshin Planning review meetings efficient and meaningful, allowing a leadership team to dispense with formatting difficulties and get right to content.

Overlooking the Need to Make Metrics Transparent and Inviting

David McCandless, in his well-trafficked TED talk says, "There's something almost quite magical about visual information. It's effortless, it literally pours in. And if you're navigating a dense information jungle, coming across a beautiful graphic or a lovely data visualization, it's a relief, it's like coming across a clearing in the jungle."[20] It's maybe a little prosaic, but Hoshin metrics tracking reports can be inviting to look at, too. It may take a few tries to get data that are truly helpful to understanding the Hoshin's health. And then it takes a few tries to get that data into a format that is palatable to the people who most have to do something with it.

In an example of how complicated a dialogue on measurement can become, and why presentation matters, senior director of quality for Advocates, Inc., a Boston-based social services organization, Carmin Quirion Wyman, explains, "When we think about something like incident reporting, at times I've brought summary data to teams, and very quickly I'm told that my data is wrong. And I'll clarify by telling them that I have the raw data as well as the 'roll up' in front of me. To this, they'll respond that the data is wrong and/or that it's misappropriated. However, when I hand out the actual spreadsheet of raw data, they'll eventually say that this information is correct, and they'll further say things like, 'Wow, that is really helpful.' So what I find is that I have to come to a discussion prepared with several visual versions of the data. For deep understanding and discussion to occur, I have to prepare to give a big picture overview, then drill down to the raw data and then move back to the summary."[21]

Summary of Effective Measurement Practices

Hoshin Planning cannot fulfill its promise for focused, responsive, and winning implementation of strategy without a consistent commitment to measuring the results of the work. There is merit to being choosy about which metrics to track and report, and to making a system for collecting data that is friendly and easy. Thanks to some decades of quality management in organizations, much is already known about how to use a minimalist approach to measurement and still get what is needed to keep improvements at the daily management level and at the enterprise-wide Hoshin level moving forward. Table 4.5 summarizes a few practices that support the use of measurement in Hoshin Planning.

Table 4.5 Some Good Measurement Practices

Some Good Practices	*Explanation*
Form metrics around good questions.	Understand what outcomes you're trying to measure and what questions you're trying to answer with your metrics.
Apply "necessary" and "sufficient" filters.	Only measure what is necessary to tell you how your Hoshin priorities are doing. At the same time, be sure to measure at least the critical few things that really will tell you if things are heading in the right direction.
Enable small, fast cycles of drafting metrics and letting new metrics emerge.	Choose some metrics and gather data in a way that can be used at weekly, monthly, quarterly Hoshin review meetings. Change metrics if it looks like they aren't right anymore.
Inventory existing measurement efforts.	Measuring can be labor intensive. Stop tracking things that are no longer relevant to the chosen direction, and reassign resources to measurement of Hoshin-critical areas.
Understand your audience.	The people who will be gathering the results and data may have the best sense of what is important to the customer and what metric will provide the best indication of whether the work being done is meeting customer needs or not. Listen to the data gatherers' observations when defining metrics and data gathering processes. In addition consider people's preferences for how they receive information from leadership and how they will present the data they gather.
Conduct Catchball on the targets.	Solicit feedback on draft metrics during Catchball meetings, those meetings set up to test the components of the plan with staff and stakeholders.
Build friendly collection processes and accessible storage for the data.	Gathering methods, frequency of checking, responsibility for checking and reporting, standard methods for putting data in storage—these things should be defined and trained on early in the Hoshin Planning process. Consider using specially designed software to manage the metrics around the Hoshin.
Assure data credibility.	Train people on the reasons a particular metric is being tracked. Never use metrics to punish or manipulate people. Where possible, hardwire the gathering of data into an e-system, reducing human error.
Take action based on the data.	Be ready to change the Hoshin plan based on what the data tell you.

With a little reflection, people in any organization will have their own lists of measuring activities that work or don't in their own particular culture and set of practices. Why bother to get the measurement piece right? Because without it, we have a hard time saying if our Hoshin did what we intended it to do. And once it does, we can show it objectively and definitively. We need the clarity of objective measureable results to close that cycle of Hoshin Planning, and move on to the next breakthrough.

Endnotes

1. Deming, Edwards W. *Out of the Crisis.* Cambridge, MA: Institute of Technology, 1982, p. 53.
2. Feiler, Bruce. "Statisticians 10, Poets 0." *New York Times*, May 18, 2014, ST1.
3. Akao, Yoji. *Hoshin Kanri: Policy Deployment for Successful TQM.* Cambridge, MA: Productivity Press, 1991, p. 158.
4. O'Neill, Julia. Merck Vaccine and Production Technical Operations, June 16, 2014, interview.
5. Isley, Malcolm. Greenville Health System, April 25, 2014, interview.
6. Frankovich, John. NV Energy, June 20, 2014, interview.
7. Rosenzweig, Phil. *The Halo Effect.* New York: Free Press, Simon & Schuster Company, 2007.
8. Cowley, Michael. Author of *Beyond Strategic Vision: Effective Corporate Action with Hoshin Planning*, May 8, 2014, interview.
9. Akao, Yoji. Ibid, p. 68.
10. Cowley, Michael. Ibid.
11. Akao, Yoji. Ibid, pp. 73–74.
12. Isley, Malcolm. Ibid.
13. Akao, Yoji. Ibid, p. 12.
14. Winters, Rich. NV Energy, May 14, 2014, interview.
15. Gold, Lois. MetLife, February 6, 2015, interview.
16. Gold, Lois. Ibid.
17. Henderson, John. Gerber Scientific, Inc., October 8, 2014, interview.
18. Gold, Lois. Ibid.
19. Block, Melissa. "Smoot, Namesake of a Unit Length, Retires." *All Things Considered.* National Public Radio. December 7, 2005.
20. McCandless, David. "David McCandless: The Beauty of Data Visualization." *Global TED*, July, 2010.
21. Quirion Wyman, Carmin. Advocates, Inc., July 30, 2014, interview.

Article 5

Building Leaders

Contents

> Leadership: Liberating people to do what is required of them in the most effective and humane way possible.[1]
>
> **Max DePree**

The Hoshin process seems to extract the best from some leaders. Shepherding a Hoshin workgroup through successfully requires facilitating a set of people who may never have worked together before in accomplishing objectives that may never have been attempted before by the organization. Hoshin work is of the most strategic importance to an organization, and when a piece of the Hoshin is assigned to someone, the assignment is accompanied by an executive mandate. So the work is highly visible, provokes change and some accompanying fear and resistance, and sometimes requires the risk of leaving one's current job responsibilities to others for the period during which the work on that particular Hoshin is active. As a leadership developmental opportunity in an organization, Hoshin Planning is hard to top.

The endnotes for this article includes just a few books from a very broad body of knowledge and scholarship on leadership. Included in this article are a few specific individual and group leadership characteristics and practices that stand out in my own work with executive teams and successful individual leaders who are living the Hoshin Planning process.

Star Qualities among Hoshin Leaders

Leadership can seem a complex subject, and so much effort has already been invested in understanding it. It's still a little mysterious when a good leader surfaces. There certainly are others, but here are four qualities good Hoshin leaders seem to share to varying degrees. And some of these qualities develop further through the process of leading Hoshin Planning work as shown below:

- An orientation toward inclusion.
- Self-knowledge and maturity.
- A realistic sense of the current mood of people.
- Enough confidence to share leadership and plan development.

An Orientation toward Inclusion

In 2010, the Power Generation leadership team at NV Energy started its second annual cycle of Hoshin Planning by creating a vision for redesigning how their workforce was organized and deployed. The work was displayed in an Affinity Diagram with more than a hundred sticky notes. The Affinity Diagram was what it always is, a singular generation and combination of ideas, unique to the specific people who were there in that specific place and under those specific conditions. Shortly after the meeting, those sticky notes were lost and not recovered.

They were busy people, and their time was valuable. Energy Supply's vice president, Kevin Geraghty, appreciates the principle of inclusion in completing the foundational work for a Hoshin plan, and he was able to embrace the loss of that first round of work as an opportunity to start over and include a key executive who had not been able to attend the first meeting. That executive turned out to be in a position that was key to this Hoshin's success down the road. So the Affinity Diagram was re-created, and with the input of the previously missing executive. Geraghty's team agreed to build the vision again, appreciating that process can be as important as content, and they trusted their leader.

Hoshin-practicing executives preside over a great deal of change. Consensus among the leadership team is essential for a Hoshin to take root. Slowing down to prioritize full participation of key people pays off in the depth of consensus the Hoshin will enjoy later.

Self-Knowledge and Maturity

Of critical importance may also be the ability to tolerate and digest unfavorable feedback from people. There is a point in the Hoshin Planning process when the Hoshin has been selected, the strategy teams formed, and a launch meeting is held. It's often at that point that some dust starts kicking up in the organization. Then at various points during the Hoshin implementation, there seems to be a resurgence of angst about the change it represents. The more obvious it becomes that the change is real, the more venomous and personal the pushback can become.

Leaders are more effective when they know their own beliefs and biases, their values, their assumptions about other people and the future, and what motivates them to do the work they are doing. The ability to conduct a personal version of Plan–Do–Check–Act (see Appendix), and then leverage that private work into a strong sense of self in public settings, bolsters a leader during tough times. A leader's beliefs show, one way or another, and one who has a good self-understanding seems to fare better under fire than one who doesn't.

Hoshin means breakthrough and newness, and people can be unsure and uneasy. Leaders who show confidence, humility, self-reflection, and a comfort level with tough feedback help generate confidence in others.

A Realistic Sense of the Current Mood of People

There is counsel provided by organizational development professionals for use during difficult and sustained organizational efforts called "Acting As If." As if the world were abundantly promising. As if as a leader you are totally optimistic. As if you have every confidence that the whole team and every individual will perform to expectations. A leader who keeps a steady positive message and faces resistance with an attitude of inquiry, positivity, and curiosity can be a great boon to a Hoshin Planning cycle. In tough situations, it's good for leaders to act as if things will work out, even if they have internal concerns.

At the same time, people shrink from a leader whose words don't reflect the difficulties they are really facing, who doesn't seem to have a clear idea of the specific challenges that will be produced by the changes Hoshin effects, or who boasts too soon of success. The founders of 37Signals, in their lean little book *Rework*, advocate candor in communicating current reality: "Don't hide behind spin or try to keep your bad news on the down low."[2] Leaders are credible when they recognize real problems, filter out ungrounded or irrational complaints, and include in their "as if" behaviors and words their acknowledgment of real challenges and specific reasons why they remain optimistic.

Confidence to Share Plan Development and Patience to See the Plan Through

Good leaders have the ability to appreciate the totality of each individual's gifts and talents and the astounding diversity that is present when those individuals all come together in a room. As important as a leader's confidence in facing uncertainty is the confidence to truly invite people to innovate and build their levels of the Hoshin plans. It engenders interest when a leader sets a direction, then lets the means for following that direction form from other people's thinking and work. Former CEO of Herman Miller, the office furniture giant, Max DePree called it, "abandoning oneself to the strengths of others."[3]

Organizational transformations often fail because leaders do not have the patience to let the plans work. In addition, there is often a mistaken expectation that goals for a one- to two-year Hoshin will always show results within a few months. Driven by short-term financial pressures, leaders can pull resources or change strategies midstream, creating a great deal of disruption and lowering the probability for success. Leaders need to have confidence, or act as if they do, that completing the plan will produce results, even if those results don't show immediately. Conducting regular reviews can boost this confidence by continually checking and reaffirming that the plan is still the right one.

A Few Hoshin Leadership Principles to Consider

Among the excellent and well-documented principles leaders can take guidance from, three stand out as particularly relevant to the practice of Hoshin

Planning in today's conditions (see the list below). Hoshin objectives take a sustained and hardworking effort over the course of a year or more. Today the longevity of service of an executive at any given organization seems to have shortened, and the almost constant improvement and increased use of information and other technologies make lots of change even more likely. These and other conditions invite leaders to bond well at the leadership level and to keep a steady focus on those transformational objectives that arise from Hoshin Planning. Strong leadership teams, not just strong leaders, fare better during times of change.

- Constancy of purpose and momentum
- Fleet thinking
- The team you're on and the team you lead

Constancy of Purpose and Momentum

The leadership team at NV Energy chose a Hoshin called *Workforce Redesign*, a fundamental change in how employees in their nine power plants would be hired, managed, trained, and assigned work. During the time of this Hoshin implementation, about two years, the vice president for human resources changed three times. Each time the VP changed, the Generation leadership team discussed whether they ought to delay their Hoshin and whether it would be impacted by a new HR executive's vision. Each time, the team recommitted to this Hoshin as essential to taking the organization to its next level of performance. The end result was a fundamental and positive change in how people at the company work. To delay based on the changes in a leadership role may have meant missing the opportunity to make the change in a way that was competitively timed.

In addition to sustaining the Hoshin vision throughout leadership changes, leaders influence the energy and momentum around the Hoshin in ways that are sometimes subtle. In her book *Leadership and the New Science*, Meg Wheatley used field theory to reinforce that "organizational behavior is influenced by the invisible."[4] Human behaviors and business outcomes are influenced by messages that make up the dialogue in an organization. Leaders need to watch out for saying one thing but sending messages that conflict with their words and actions. For example, one healthcare leadership team promoted safety awareness and a faithful reporting of "near misses," but continued to reward managers who provided clean, incident-free monthly reports. We can't always spot a conflicting force like this early, but we can see the effects of it over time, in people's behavior.

There are all kinds of messages moving through organizations, fields of thinking that are often not formal. Wheatley suggests we pay attention to the "fields" we form. People will sense when a leader has shifted priority from the Hoshin onto other things and will respond to the Hoshin work with lackluster commitment. Changes in a leader's commitment, contradictions in her or his message, and subtle shifts in focus and priority are not only felt; they are also responded to by people in the organization.

Fleet Thinking

Machu Picchu was designed and built around 1450 A.D. The Inca managed to keep it more or less a secret from the outside world until Hiram Bingham announced its historical significance in 1911—almost 500 years later. Perhaps one reason the Inca achieved this level of propriety and secrecy may be that the ninth Incan leader, Pachacuti, conquered or partnered with a number of disparate tribes across the west coast of South America, united them, demanded they speak a shared language (Quechua), built roads to link all the tribes' areas, and declared that together they were the Incan Empire. He led them as one fleet with one shared purpose—to grow their wealth and power and in the end, to fight off the Spanish. By the time the European Spanish took over in the 1530s, the empire was capable of sustaining a shared effort. The last Incan ruler, Tupac Amaru, was executed by the Spanish leader, Toledo, but it was another 500 years before Machu Picchu was found by the European conquerers—a long time to keep a shared secret.[5]

More recently, Julia O'Neill, a Hoshin program leader at Merck, helped pull together five independent departments related to vaccine manufacturing to form a team called Vaccine Product Technical Operations (VAPTO). This merging was done prior to starting Hoshin Planning. As a result, the Hoshin objective chosen was one that crossed all five departments and was designed to raise the entire entity to its next level of performance. To lead cross-platform, cross-department, cross-functionally, the leaders of all the parts must be involved in building shared objectives. Adjusting formal lines of reporting that support the development of fleet thinking may help that process.

Executive level leaders can consider whether it makes sense to start preforming what might be the leadership team that will be the right team once the Hoshin becomes the new way of doing business. It makes sense to include those leaders in the process of selecting and deploying the Hoshin. If the Hoshin vision includes separate parts of the business coming together, leaders might start pulling together those disparate leaders as part of the Hoshin development process.

The Team You're On and the Team You Lead

An enterprise-wide Hoshin means that every department leader must understand that she or he takes direction from the central body of decision makers. Author of *The Advantage: Why Organizational Health Trumps Everything Else in Business*, Patrick Lencioni writes, "The only way for a leader to establish a collective mentality on a team is by ensuring that all members place a higher priority on the team they're a member of than the team they lead in their departments."[6] When this doesn't happen, Lencioni likens it to the U.S. Congress, "just a place where people come together to lobby for their constituents." When the organization commits to a Hoshin, that Hoshin is resourced before individual business units receive resources, and its priority over personal or unit objectives is understood.

Some competitive representation from the parts of the system can be good, something like how Doris Kearns Goodwin describes Abraham Lincoln's Cabinet in her excellent leadership book, *Team of Rivals*.[7] Profoundly different paradigms, developed from very different professional and personal paths, can generate a great deal of creative energy on a leadership team, if the resulting various views are respected and challenged equally. Further, Lincoln's senior level Cabinet included William Seward, Salmon Chase, and Edward Bates, three men who ran against him in the presidential election, lost, and in at least the case of Chase continued to want Lincoln's job. Over time, the natural tension between those men produced well-vetted and bold decisions that achieved two Hoshins of spectacular scope: the preservation of the first democratic union of states and the abolition of slavery.

This kind of reflective and confrontational shared decision making differs substantially from the congress-like advocacy that results from leaders not identifying sufficiently with goals shared at the highest level of the organization. Lincoln's Cabinet practiced what Chris Argyris popularly called inquiry, not advocacy. In an inquiry model, regional or departmental leaders come to the table with a shared commitment to common goals and then question and shape decisions related to those goals through the filters of their own constituencies, but with an openness to learning and changing their minds.

Kevin Geraghty at NV Energy says it well in reference to his team of plant directors. "There weren't any bad characters, but you could see that if people say 'If I optimize my stuff, that'll be good for the company,' but it won't really be OK, right? How you fit in and how you work and what we need from each leader changed dramatically when we started to build one central Hoshin plan. I said to my team early on 'You're going to feel unconnected if you don't change to a shared thinking, and I don't want that for you.' Hoshin is a unifying method around this core team thinking, where you agree that plant A might have to take a hit for plant B. There's a bigger bang for the customer, if we work on common goals."[8]

Some Essential Hoshin Leadership Team Practices

Perhaps the quality of a leadership team is most revealed by how well it makes decisions. In addition, how a team prioritizes, delivers its message, and learns together shape its success in leading its organization. Attention to these four things, summarized below, during the course of the Hoshin work can bring strength, growth, and clarity to individual members of the team and to the team as a whole.

- Making decisions
- Prioritizing
- Messaging
- Learning

Making Decisions

Turning interesting strategic dialogue into meaningful guidance to an organization is one of the essential acts of a leader. Leadership work at meetings is not over when everyone has provided their best and most interesting idea. Sometimes a kind of academic orientation dominates leadership meetings, where leaders are steeped in the practiced ability to listen efficiently, not necessarily thoroughly and with reflection, provide an "A" answer, and move on to the next topic. But the sustained work of talking together to create shared meaning, reworking and exploring ideas, and arriving at a decision together is work that has to be done in leadership teams practicing Hoshin Planning.

Leadership teams might follow a simple set of questions when they're facing a decision: What is the decision to be made? Who will make it? What method will be used to make the decision? What information is needed to make the decision well? How will the decision be communicated? There are four basic ways to make a decision, and high-functioning leadership teams become skilled in choosing which one works best for which situation (Table 5.1).

Table 5.1 Decision Types

Choice	*Description*	*Advantages*	*Disadvantages*
Consensus	A process by which everyone is heard, and every leader has the power to veto a proposal. Used when total buy-in is essential.	Involves extensive dialogue and a greater understanding of competing views.	Time consuming. Can produce poor quality decisions, if the data are flawed or people's interpersonal skills aren't very good.
Voting	Used when the list of options is long and the time to choose is short.	Is a democratic process. Creates a measurable show of how much support there is for the decision.	Some win, and some lose. Often involves limited dialogue.
Directive	One person decides. All follow. Used when one person is the expert and/or is willing to have exclusive accountability for the outcome.	Is efficient. Produces clear accountability.	Possible lack of buy-in. Possible exclusion of high-quality input.
Compromise	A negotiated agreement. Used when consensus is not possible or important, or positions are very polarized.	Generates some discussion. Builds a contract that represents most interests.	Is adversarial and divisive. Everyone wins. Everyone loses.

Prioritizing

Hoshin leadership teams must practice the difficult task of prioritizing. This usually requires a reduction of strategic level activities, and may involve feelings of frustration and loss for leaders. The loss is sometimes temporary, if the multiyear sequencing of the Hoshin objectives shows how a member of the leadership team's "pet" Hoshin will be featured a little further down the road. Or a loss of an organizational activity may be permanent, because it is outside the path the C-suite has chosen for future growth. The reluctance to remove things from an organizational plate seems natural. Leaders may feel irresponsible not taking an opportunity that looks good or fixing a problem that looks big, regardless of its fit with mission, market strategy, and organizational skills. Further, those who must send the reduction message up or down a level in the organization invite particular compassion. It's no small thing to suggest to someone that his or her project may not be the best use of resources at this time.

Who doesn't want it all, or believe during moments of achievement and confidence that not just anything, but *everything* is possible? But leaders must be able to make work that is no longer supportive of a focused future vision go away. Hoshin Planning provides an opportunity for exquisite focus. How well leaders can say "no" will determine what resources—dollars, people's energy and time—are available for Hoshin and other strategically relevant activities.

Messaging

Leaders must speak with one voice, no matter what. Shared scripts and some practice help leaders consistently project the same clear message to the organization. Article 6 on change offers more on specific content for those scripts and messages. In addition to full agreement among leaders on the few key priorities and pieces of information they all will express, leaders need to be ready for a long cycle of repetition. Key messages must attend every meeting, Hoshin-related and other, as a kind of mantra that people begin to expect and to which they attach meaning.

Peer Leadership Learning

Each leader has some responsibility for helping advance the growth of their fellow leaders, a kind of developmental obligation to peers at their own

Table 5.2 Leader-to-Leader Support

Show up.	Stand with her or him when he or she has to deliver a tough message.
Be a trustworthy listener.	Periodically provide a confidential ear for impolitic or less rational thinking, without judgment or repercussions.
Say "yes" with conviction.	Follow through on commitments and deliverables promised.
Say "no" well.	Be timely, candid, and reasonable when another leader asks for something you cannot provide or believe would be imprudent to do.

leadership level. Each leadership team member developing the ability to make and keep commitments to other leaders produces a deeper level of accountability. Lencioni writes "Peer-to-peer accountability is the primary and most effective source of accountability on a leadership team." The most critical and final decisions get made at the very top of organizations. When leaders can count on each other to do what they say, that kind of stability and confidence shapes cultures and spreads a correct understanding of accountability among the people they lead.

In addition, there are acts of social generosity that can help peer leaders leverage the value and ease the strain of their shared learning. Table 5.2 shows a few practices for supporting a fellow leader. When all members of a leadership team are direct and challenging, and also compassionate, their ability to get big things done for and with the organization increases.

Leadership Team Self-Evaluation

Leadership teams ought to periodically check in on their own leadership performance. Figure 5.1 provides a tool leaders can use to check their health as a team. Done semiannually or annually as part of a leadership retreat, such a tool can serve as a start to dialogue around how the team can improve its leadership of the organization and areas around which they can be proud and confident. Executive teams might then ask leadership teams at the next levels in the organization to adopt the evaluation practice, too.

Trust			
Members of the team are closed, withholding their responses to events or information.	1 2 3 4 5 6 7 8 9 10		Members of the team can be genuinely vulnerable during meetings, open about mistakes, and trust the team to hear them.
Conflict			
Conflict is avoided, or time is not made available to understand sources of conflict. Differences in perspectives and opinions are not explored.	1 2 3 4 5 6 7 8 9 10		Conflict is invited and managed. Sources of conflict are understood. Ways of responding to conflict are productive. Differences are used to improve the quality of decisions.
Decisions			
Leaders have different understandings of what has been agreed to. Leaders act independently of team decisions. Important decisions don't get made.	1 2 3 4 5 6 7 8 9 10		Decisions made at meetings are specific and documented. All leadership team members act in full accordance with decisions made by the team.
Accountability			
Leaders are not held accountable by each other or by the organization, for agreements, behaviors, or results.	1 2 3 4 5 6 7 8 9 10		Leaders hold each other accountable. Leaders hold their reports accountable for both behaviors and results.
Priorities			
Goals owned by the leadership team are not clear or public. Leaders privately prioritize department and unit goals over organizational goals.	1 2 3 4 5 6 7 8 9 10		Leadership teams' shared goals are clear and public. Leaders make decisions as if organizational shared goals take priority over individual department or unit goals.

Figure 5.1 Leadership team evaluation.

Hoshin as a Leader Builder

Hoshin Planning helps create good leaders. People who lead strategy work-groups or take the role of Hoshin program manager generally come out on the other side of that work more skilled. In addition, leadership teams grow together in their ability to make consensus decisions, prioritize and focus their organization's work, and think about each other less as managers of another silo than as shared custodians of the organization's success. "It wasn't just the output of the Hoshin itself, but the act of doing the Hoshin together that was really powerful. Working together to develop the Hoshin plan, we really gelled as a leadership team,"[9] says Merck's Julia O'Neill. It's a big step on the developmental path of an executive team to come together with colleagues and successfully guide an organization through the focused transformation Hoshin Planning can deliver.

Endnotes

1. DePree, Max. *Leadership is an Art.* New York: Dell Publishing, 1989, p. 1.
2. Fried, Jason, and Heinemeier Hansson, David. *Rework.* New York: Crown Business, 2010.
3. DePree, Max. Ibid. p. xvi.
4. Wheatley, Margaret J. *Leadership and the New Science.* San Francisco: Berrett-Koehler, 2006, p. 56.
5. Heaney, Christopher. *Cradle of Gold: The Story of Hiram Bingham, A Real-Life Indiana Jones, and the Search for Machu Picchu.* New York: Palgrave Macmillan, 2010.
6. Lencioni, Patrick. *The Advantage: Why Organizational Health Trumps Everything Else in Business.* San Francisco, California: Jossey-Bass, 2012.
7. Kearns Goodwin, Doris. *Team of Rivals: The Political Genius of Abraham Lincoln.* New York: Simon & Schuster, 2005.
8. Geraghty, Kevin. NV Energy, June 6, 2015, interview.
9. O'Neill, Julia. Merck VAPTO, June 16, 2015, interview.

Article 6

Hoshin Disruption

Contents

> When the Wind is fair and the planks of the vessel sound, we may safely trust everything to the management of professional Mariners; but in a Tempest and on board a crazy Bark, all must contribute their Quota of Exertion. In the present agitations of the public mind, everyone ought to consider his intellectual faculties as in a state of requisition.[1]
>
> **Samuel Taylor Coleridge**
> *18th century poet*

Coleridge was generating enthusiasm for a revolution, reminding listeners when he spoke this poem during a lecture that every person would be pressed into service to make the change. Hoshin Planning invites an organization to commit to a substantial breakthrough in performance, a revolution in thinking and practice. Achieving the breakthrough involves an organization doing things substantially different from how things are currently done. How well we help each other through the technical, procedural, and behavioral changes required has a striking effect on the pace of the Hoshin implementation and on how healthily people internalize the change. Hoshin creates a significant disequilibrium, a level of disruption that allows for transformational change in an organization.

Changing a Familiar Organizational Habitat

Biologists use the term "background extinction" to describe the gradual, incremental loss of species that the Earth experiences on an ongoing basis, nature's constant adjustments to change. This level of extinction is contrasted with mass extinction, a kind of steep crash of loss, center stage in evolutionary history. Perhaps process improvement and Hoshin Planning sit in similar relation to each other as "background" and "mass" extinction do. The focused and relatively fast pace of strategic level change that goes with Hoshin Planning can create a kind of mass extinction of practices in organizations. Growth and loss move from the background level produced by routine process improvement changes to the mass level produced by a fundamental change in the systems and practices of the organization. There are organizational habitats that will become extinct as the Hoshin is fully implemented.

Some people will let go of old practices and lean into new ones with some gusto. Others seem to experience loss and sadness, sometimes to a debilitating extent. The change associated with Hoshin Planning may call for leaders to witness, mitigate, and sometimes tolerate the discomfort of the people they lead. Executives and managers need to sift their decisions and actions through a mesh of compassion, as people's existing ways of doing things are changed, and yet they must still hold people accountable for implementing that change. This is hard leadership work.

Hoshin Characteristics

- Requires a fundamental change in one or more of the organization's systems.
- Involves most parts of the organization in achieving it.
- Will take the organization to its next level of performance—a step change.

A strategic objective registers as a Hoshin, with the associated level of change, if it has at least three characteristics. To bring along change of this magnitude may require leaders to help migrate the organizational community from a comfortable habitat to a temporarily uncomfortable new one.

Professional Identity and Change

In addition, to embrace a Hoshin change might involve creating some temporary distance with a person's professional community and history. The more stature someone has in a profession, the more likely it is he or she will be seen as representative of that profession's values and practices. To embrace a change that calls a profession's previous practices into question is to risk a person's alienation from the profession's community. Among astronomers in the sixteenth century, Copernicus was the first one to declare in writing that the Earth revolved around the Sun, and not vice versa. Some of what he got for that brave and visionary move was the rejection and intellectual refutation of that century's astronomical darling and his peer scientist, Tycho Brahe, and the slow but steady exile from a religious community that was hanging on doggedly to the Earth-centric story it favored.

The philosopher and Texas Tech University professor Costica Bradatan offers a description of exile that sheds a little gentle light on what people distanced from their professional "world" might feel.

> To live is to sink roots. Life is possible only to the extent that you find a place hospitable enough to receive you and allow you to settle down. What follows is a sort of symbiosis: just as you grow into the world, the world grows into you. Its culture shapes the way you think, its customs structure you as a social being. Who you ultimately are is determined to an important degree by the vast web of entanglements at "home."[2]

Some strategic changes in organizations put pressure on a person's professional "home." As an example, if someone spends a couple of decades as a power plant supervisor who on arriving at work each day checks a clipboard, pencils a few notes, assigns work to his junior tradesmen according to his own best judgment of what needs to be done, and then hangs the clipboard back on its hook, his practice may feel good and familiar to him.

If he's been working that way a long time, and the plant is running well, he's probably earned a good deal of respect from his peers for his knowledge and skill. It's possible he takes pride in being able to do it all "on the fly," and he may believe that he can simply do it better than some analyst could. And he may be right. He's achieved a kind of "Look, no hands!" level of personal performance. But his success is person-based, not scalable or sustainable. What happens when he leaves?

If the people in the C-suite tell him he now has to turn on a computer first thing each day and read some irrelevant-looking numbers about how the plant performed over the last month and year, then assign work according to what he planned (or someone else did) a couple of weeks ago, and then record a bunch of things at the end of the day using his hands at a skill he doesn't value—typing—his professional sense of value may be stressed. A significant source of his stature with his peers may feel like it's being taken away. And some of the people working for him may seem to hate the change to different ways of getting work done, to boot, so they may call him a traitor to their shared craft. All of this tension and loss may push him to fight about it with anyone who will listen, and feel righteous in what he believes is the defense of his workers, his plant, and his profession.

If you substitute a field engineer, nurse, social worker, electrician, teacher and so on for the plant supervisor, you may see a similar conflict between change and the preservation of professional community. If this leaves the impression that only crafts, trades, or frontline professionals are at risk for this sense of alienation, no. Financiers, software programmers, physicians, quality managers, small business owners, and management consultants all become attached to the core principles and practices in their professional ken. Many of us identify strongly with our work and feel like ambassadors for our own professional communities. This type of pride is wonderful and even as it can be a barrier to growth and change, it is worth nurturing when we meddle with people's day-to-day work, as Hoshin Planning often does.

Climbing Change

It may be helpful to think of the change associated with Hoshin Planning as a series of ascents to plateaus. The trail climbs, winds, and at times is lost and must be retraced. Change is iterative, and people in an organization move through it at different speeds. Which plateau someone is standing on

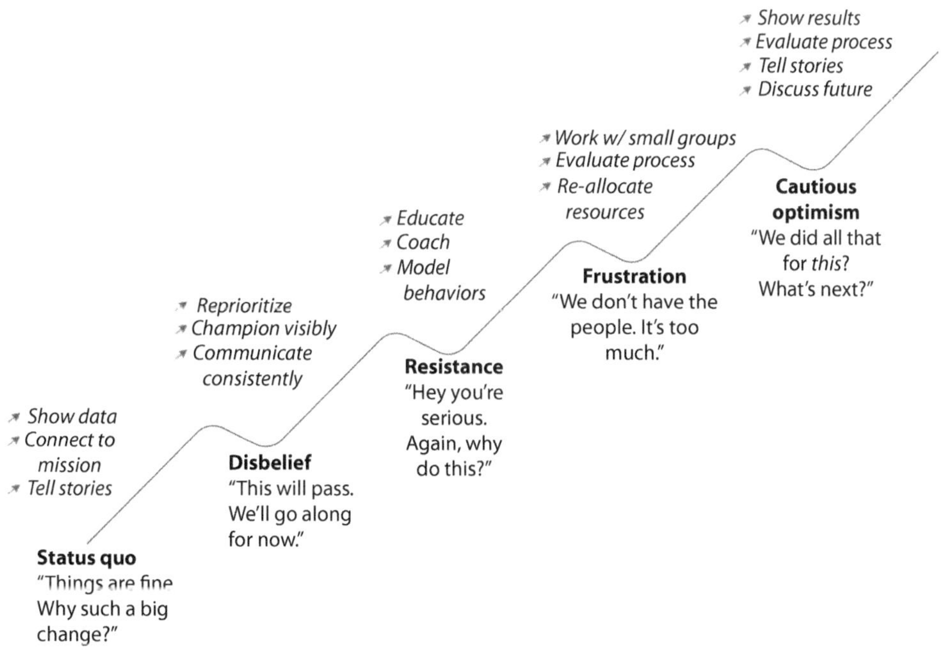

Figure 6.1 Hoshin change platforms.

can sometimes be determined by the things he or she says. Figure 6.1 shows a picture of how this ascent can look as Hoshin Planning moves through the organization.

The supporting activities next to the arrows in Figure 6.1 aren't especially complicated… but it's hard work to do them consistently and over a longish period of time. It can help to think about how far along the organization is emotionally. The best leaders stay agile, trying things until something works, and moving back and forth between the plateaus depending on who they are coaching at any given moment. It can look tempting to charge up the slopes at the pace of the fastest climber. It's worth trying to find out how fast you can go, but it can happen that a group of people are left a couple of plateaus behind, building encampments and brooding on ways to block the people behind them from getting any further.

Resistance and Individual Power

An emotional and vocal representative of a minority view can stop a leader in her or his tracks during the presentation of months of Hoshin-related planning work. Individuals resisting the change presented by the Hoshin can filibuster progress, if they get airtime (a good thing) and are allowed to rewrite the story of the Hoshin through filters of fear, distorted facts, and unproductive biases (a bad thing). Vocal resistors can feed a kind of organizational auto-immunity to the change the Hoshin represents. And when the whiff of change is in the air, organizations' chat mechanisms seem to kick into overdrive. All of those independent stories can be confusing to people and can make healthy change more difficult. The Czech social scientist Karl W. Deutsch used the word "jamming" to describe the crowding of competing messages into people's thought space. He wrote in the 1960s that an individual could possess significant power by refusing to learn, jamming incoming messages with his or her own long-held internal ones. Deutsch's often-repeated quote about resistance is, "To have power means not to have to give in, and to force the environment or the other person to do so. Power in this narrow sense is the priority of output over intake, the ability to talk instead of listen. In a sense, it is the ability to afford not to learn."[3]

A resolute resistance to learning, and the isolation that comes with it, can create insensitivity to the present. Duetsch describes the person stuck in an "I won't learn" mode as, "driven like a bullet or torpedo, wholly by [the] past." New information, changes in the environment, and the insight available through making a fresh way with others are lost to that person.

Much of what we do in Hoshin Planning to move an organization closer to its objectives is highly nuanced. If leaders' invitations to change are too ambiguous, people can dismiss them, but if they're too direct or forceful, people can dismiss the messenger him- or herself. We're at once working to step up the anxiety level and to create a climate that is emotionally safe for making the change. It's right to saturate the organizational environment with knowledge and skill building mechanisms without overloading people. What a dance Hoshin leaders must do among these forces.

Conscious Disruption

The social psychologist Edgar Schein describes organizational culture as deep, broad, and stable. Hoshin Planning generally disrupts the stability, producing for some anxiety. The anxiety, for many, will be related to their ability to learn what they must to make the change and the implications of not doing so. Schein outlines five sources of learning anxiety in his book *The Corporate Culture Survival Guide*.[4]

1. Fear of loss of power or position
2. Fear of temporary incompetence
3. Fear of punishment for incompetence
4. Fear of loss of personal identity
5. Fear of loss of group membership

People might have to get pretty scared to enter a path of transformational learning. And perhaps it's that profound, transformative level of individual learning that lies beneath fundamental change in an organization. In a 2002 interview Schein said, "I believe that all learning is fundamentally coercive, because you either have no choice, as is the case with children, or it is painful to replace something that is already there with some new learning."[5] Change can be scary.

A leader ought to consistently and compassionately let people know that the direction chosen through the Hoshin Planning process is the one the organization *will* take. Some current practices will be disturbed and changed, and some people won't like it.

If we believe Niccoló Machiavelli, "People will live quietly and peacefully so long as their old ways of life are undisturbed." A good Hoshin implementation is likely to produce breakthrough thinking and ideas that, when implemented, will necessitate the learning and change, in some cases profound, of most people in the organization. Leaders can try to break up some old and unhelpful ways of thinking and encourage people to think more diagnostically about their industry, organization, and profession. Such leadership care can help increase peoples' understanding of their own patterns of thought by listening and talking consciously and consistently, and by being absolutely clear about the reasons for the Hoshin.

Actively Advancing the Change

It's best not to wander up the path of plateaus in Figure 6.1 too passively, trusting the levels of change will be gained without specific and planned effort. There are practices leaders can employ at each stage of Hoshin Planning to shepherd the organization through a healthier change. Some specific practices are outlined in Figure 6.2.

	Some Common Organizational Behaviors	Leadership Actions
Select Hoshin	• *Satisfaction at being asked for input* • *Focused attention on leaders* • *Enthusiasm or "ho hum" for selected Hoshin* • *Wariness of new "flavor of the month"*	• Thank those who provide input. • Be visible/consistent as leadership team. • Accept and explore disagreement. • Describe improved state that is possible. • Show evidence that change is needed.
Align/Deploy Plan	• *Anxiety about change and loss* • *Some territorialism* • *Concerns about resources and time* • *Broad interest in providing input*	• Listen without judgment. • Explain what might change. • Acknowledge the losses that will occur. • Describe risks and provide assurance. • Coach on the use of planning methods.
Implement Plan	• *Individuals unsure what is expected* • *Inconsistent acceptance of methods* • *Frustration in resolving roadblocks* • *Loss of interest*	• Provide clear roles for people to fill. • Provide accurate, accessible plan details. • Educate on PDCA methods and tools. • Communicate mission, metrics, reasons. • Coach and encourage.
Conduct Reviews	• *Anxiety about using metrics to evaluate* • *Inconsistent skills in cause analysis* • *Fear of criticism*	• Create thoughtful measures of success. • Build skills in giving data-based feedback. • Reward successes.

Figure 6.2 Change and leadership actions.

In addition to these ideas, there are five overarching practices shown in the list that follows that may support people in making the most of the Hoshin change. These practices can be built into the communication and change plans that are essential to include in an organization's Hoshin Planning process.

- Make a case that Hoshin is needed for organizational survival.
- Build bridges from existing practices to new ones.
- Ask for and receive input.
- Give information.
- Manage information overload.

Making a Compelling Case for Hoshin as Survival

It's a good idea to explain the economic realities of the organization's marketplace. Teaching people about the conditions in which the organization must succeed can help build credibility for the reasons the Hoshin is being implemented and the changes it will involve. A company in the power industry in the early 2000s might have included in its early message around the Hoshin a reminder that Congress was likely to pass a carbon tax soon and that regulators might not be allowing a rate increase to be passed on to consumers (inefficiencies could not be paid for with higher revenue from existing customers). Making fundamental changes to how they delivered power would happen, or they would go out of business.

This kind of explanation might make people nervous, but it respects their intelligence and provides data-based talking points for leaders at all levels of the organization to use when discussing the changes. Edgar Schein said in the previously mentioned interview, "We should focus on the validity of what it is we're trying to teach."[6] If we share good quality information when people need information, and make learning accessible, it might feel less "coercive" to insist that people leave their comfort zones and make a change.

Bridging

Hoshin plans often do not get built on the blank slate of a startup company, but rather in organizations with an existing body of experience and expertise. People who are working to implement the Hoshin already know a great deal and have experienced things that may be similar to what the Hoshin presents or that are very different but informative in some other way. It can help to connect the Hoshin work concretely to things that people already know, to build bridges to the new information from the existing. To do this is to respect an individual's experience as well as promote organizational learning.

In addition, leaders can use existing language to describe future states, letting old names for things linger for a while in transition, for example. And there is merit to asking people to describe what did and didn't work about the old state of things and about their previous experiences of major change. Leaders can link those pieces of input to what people can expect to see as the Hoshin Planning process unfolds, explaining what might be the same and what will be different.

Asking for and Receiving Input

Hoshin Planning leaders may want to put new or renewed focus on observing and listening to the people in the organization. Communication of an accidental kind is not sufficient to support large organizational change. C-suite leaders gain considerable advantage by making concrete communication and change management plans part of the overall infrastructure of the Hoshin Planning process. Soliciting input from people is an essential part of those plans.

Voice versus Vote

It's relatively easy to provide just about everyone in the organization with a voice as part of Hoshin Planning. It would be impractical, though, to provide everyone with a vote on critical decisions. In fact, most people in organizations will not have a vote when it comes to the direction of the organization. Making it clear when people are being asked only for their voice and when they will be provided with a vote can save some trouble later. Most of us seem to accept either invitation, but don't like being led or allowed to believe we have a vote when we only have a voice.

Surveying as an Inquiry Tool

Simple surveys are a fine way to solicit input from employees, if leaders communicate clearly what will and won't be done with what people offer in their responses. Leaders can reject ideas, but it's best to do it with good rationale and within a reasonable amount of time after having received survey results. The list below offers a few sample questions for checking the wind direction partway through a Hoshin-generated change.

- Do you understand the reasons for the change?
- Is it clear to you how the organization will be stronger/more profitable/better positioned after the change?
- Do you understand your role in making the change and what your role will be after the change?
- How confident are you in fulfilling those roles?
- What support do you need to help make the change?

Paying Attention

We don't have to be professional psychologists to thoughtfully tune in to how people talk and behave during meetings, at Hoshin presentations, in the hallways, and through their typed words in the company chat room.

More than likely, as leaders begin to talk about the Hoshin in meetings, stories will begin to be told. Facts will mix with fiction created by anxiety around change, and personal journeys toward support or resistance toward the Hoshin will already be in motion. Leaders can listen to these stories, notice what is being accepted as fact, and identify real-time opportunities to contribute to the story that is forming. This kind of influencing should be built into a carefully laid out communication plan.

Superimposing a clean controlled story over the one forming in the organization on its own doesn't seem to work as well as scraping away a bit of misinformation and attaching something factual to where it was. Like sanding a painted surface first to remove the loose chips before laying another coat on, some new story paint will stick to what is already being told and some will adhere to the original raw surface. It's good to leave in whatever is good in the organically forming story.

To make this happen, leaders have to create opportunities to observe, and the communication plan should outline those opportunities. Being an effective observer takes time and the ability to achieve a certain degree of invisibility in a room. Most leaders arrive at the executive position with a couple of decades of conditioning to talk, present, direct, decide, and otherwise "manage" a situation. To observe actively, meaning not just sit in the back of the room and audit, takes some practice. Eye contact, sitting right in the space with everyone else, and, only after some focused time, paraphrasing back what is heard and seen, are good places to start.

In his excellent book, *The Skilled Facilitator*, Roger Schwarz invites us to take clear pauses between what we observe, what we infer from it, and how we decide if and when to intervene.[7] Being a good leader during the Hoshin change includes applying discipline to listening practices, a dedicated effort to observing and reflecting. An executive is also a facilitator at times and must take care not to be hasty in attributing resistant behaviors to the Hoshin change when what she or he is seeing might just be the organizational system working itself through anxiety and loss as old practices end and in putting words to those feelings. It's worth at first listening without judgment and making inferences inclusively, inviting people to share in a dialogue about what the leader is hearing and observing.

Sharing Information

"Communication" comes up in almost every Hoshin Planning Affinity Diagram (see Appendix for more on this tool) as a header card. It is so predictable that it can sometimes be relegated to some vague level of priority, something understood to be absolutely essential and yet somehow hard to make actionable.

Leaders can assign the communication category of work to a strategy owner, as it would with any other prioritized element from the initial visioning exercise, or they can take it on as a shared action item at the leadership meetings. Communication continues to show up in early Hoshin Planning brainstorming exercises, because it reflects a legitimate appetite for information, one that is often unmet.

It might help to consider a version of that "Communication" category that uses the word "Information," to see if that generates more meaningful dialogue about how to help the organization communicate about the change that Hoshin Planning will require. Information ought to be considered a continuously generated, constantly shared, and transparently delivered resource. What people seem to crave is not only an opportunity to interact, but also, and perhaps more so, accurate, timely, and relevant information about the work they are expected to do.

So, in addition to considering the various, well-practiced and well-documented means for communicating with people in an organization, we need to care about the *content* of those communication means and the manner in which people absorb and digest that content. Two good pieces of information to include are a description of what the organization will look like after the Hoshin and an explanation of what the Hoshin pathway will look like.

Describing the Post-Hoshin World

Providing a picture connecting how the organization is now and how it's expected to look in the future helps people adjust to what will be different. "A lot of the Hoshin process was about helping people understand that we had to figure out our identity and destiny. As we worked through it, people really understood the value of thinking about who we are,"[8] says Julia O'Neill, Hoshin program manager for Merck Vaccine and Production Technical Operations (VAPTO). Leaders use pictures of a future state with Hoshin teams and stakeholders outside the group to generate discussion and build support for the change.

Organizational charts, "before and after" lists of services, new macro-level flowcharts of how the business will run after the Hoshin implementation—these describe what the organization will be like after the Hoshin and give people something in which to find a place for themselves. It's important to distinguish boundaries between services currently provided or processes currently done, those that will be redirected or changed, and those that may need to be developed as part of the Hoshin Planning work. A picture is a good bridging tool, helping the organization compare its current identity to its post-Hoshin one.

Explaining the Specific Hoshin Pathway

In addition, the Merck VAPTO leadership team took time to identify the expected sequence of Hoshin implementations over the next several years, acknowledging that things could look different at each turn in the path when they arrived there. The team created the image in Figure 6.3 to show

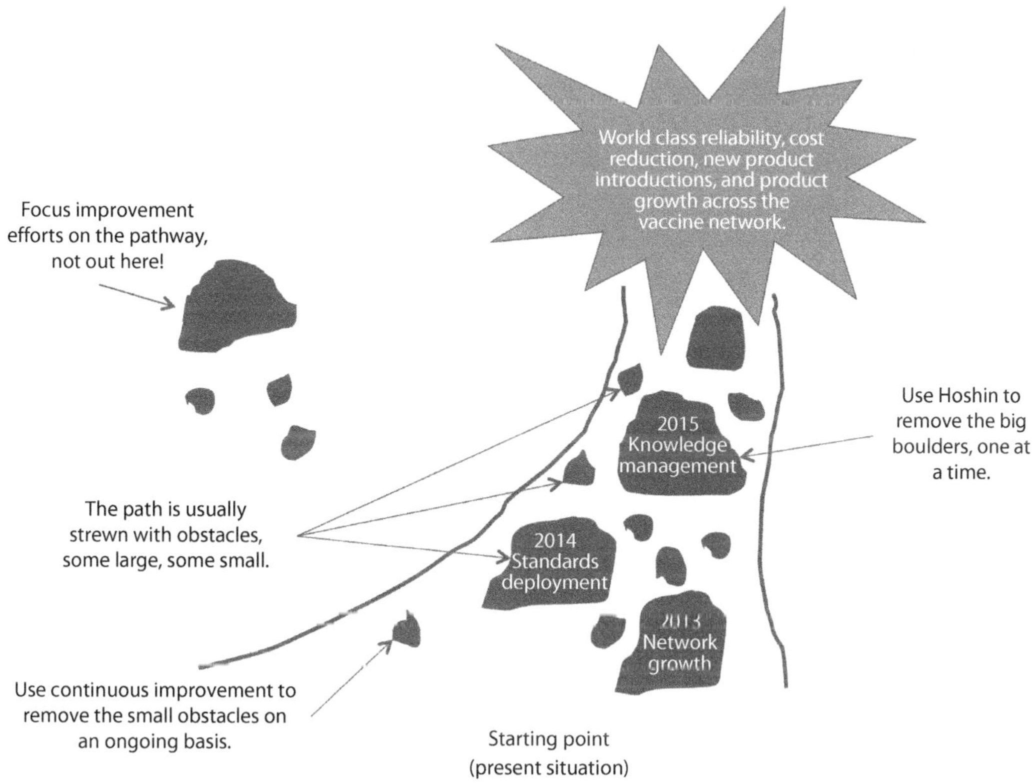

Figure 6.3 Merck VAPTO Hoshin sequencing. (From Cowley, M. and Domb, E., *Beyond Strategic Vision: Effective Corporate Action with Hoshin Planning*, Routledge, New York, 2011.)

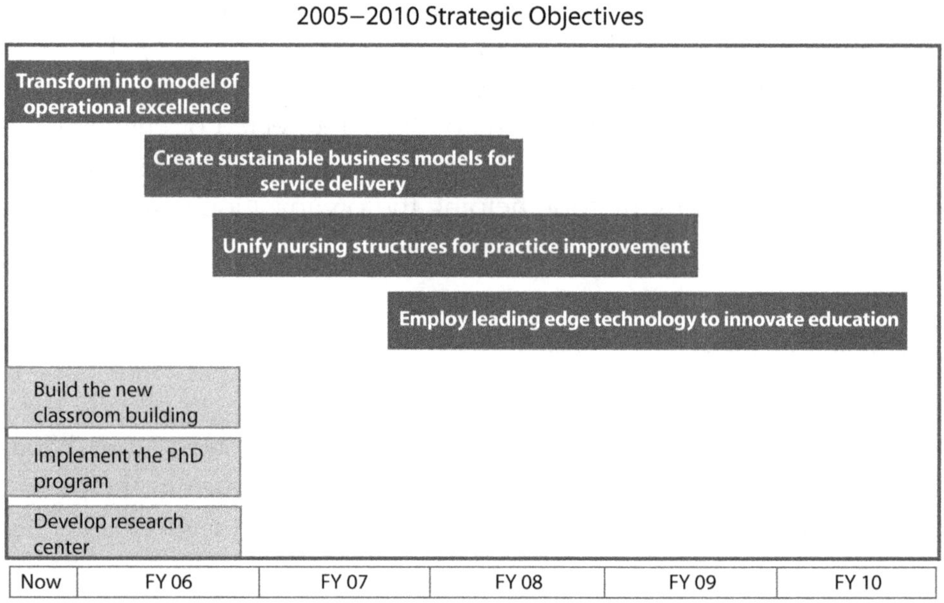

Figure 6.4 Gantt sequencing of Hoshins.

how the successive Hoshins would unfold over the course of three to five years. This image was developed for use in a Hoshin Planning context by authors Mike Cowley and Ellen Domb in their book *Beyond Strategic Vision: Effective Corporate Action with Hoshin Planning*.[9]

Leaders may also use the Gantt chart to show how the Hoshins are expected to lay out over time. Figure 6.4 shows an example from a nursing school. The three gray items at the lower left are not Hoshins, but rather large projects that were already begun and that were expected to impact resources as the first year Hoshin was deployed. Together, the bars on this chart give an audience a quick picture of the strategic priorities of the organization three to five years out.

Managing Information Overload

Social scientists say our capacity to hear and digest information is limited. How much leaders say, overlapped with how much people take in, equals the only information that is actually transmitted. In fact, as we increase the amount of information to people, brain science says they will begin to make more and more errors. The social psychologist George Miller explained,

"When we increase the amount of input information, the transmitted information will increase at first and will eventually level off at some asymptotic value."[10] Asymptotic, in math, describes a limiting function. So we get more of what is said to a certain point, and then the quality of our intake declines. We just can't absorb more than a few things at a time when the stakes are high and what is happening is substantially new to us. Miller calls the amount we can really digest our "channel capacity."

For the purposes of Hoshin-related change, it may be enough to consider that we have a finite capacity for digesting and making meaning from more than a few pieces of discrete information at a time. For example, "the humidity is high," "the ground temperature is low," "the wind is picking up," and "the barometric pressure is over" could be a great deal to hear at once. Using the abstraction "looks like rain" may be a better way of getting the messaging job done.

"Abstracting" and Stories

We abstract to get a large amount of detail into a small messaging space, to get to the essence of things. Our minds turn bits of information into larger chunks and recode minor pieces of information into broader, representative ideas. Probably organizing a message into chunks from bits helps expand a listener's capacity and stretch any information bottleneck. This chunking can look like a story.

Among many of the successful executives I work with there is a practice of storytelling that personalizes and strengthens the message they deliver. Much good research has been woven skillfully into Jonathan Gottschall's inviting book, *The Storytelling Animal*.[11] He shows that our mind is designed to detect patterns and meaning from what we experience and the facts we receive, but that where there are gaps in that information, people tend to invent things. The human brain is wired to dislike uncertainty and confusion. This makes us a bit vulnerable to filling things in, making things up, when the "real" facts are lean.

The Hoshin plan should include a scheduled set of specific communication activities. In addition to the conscious two-way messaging and information gathering work, leaders ought to create organized ways for people to paraphrase what they've heard, providing C-suite representatives an opportunity to correct misunderstandings and retell the story.

Stacking Meaning

Like getting the first step of reasoning wrong, but using it for the rest of the problem and being sure it's correct, people resisting change may take an old belief and hear every story through that. Given enough story rope, they can temporarily hang your Hoshin. If leaders don't help the story unfold accurately, they put the change effort at risk for what is called the "anti-story," a story that works to dislodge the original story, regardless of its accuracy.

Leaders can in some way chaperone what neuroscientist Michael Gazzaniga describes as the "interpreter" in each person.[12] This is the set of circuitry in the human mind that brings order and logic to what a person hears and experiences, turning it into a story he or she can use to make sense of the input and communicate intentions and plans. Each of our interpreters passes that meaning through the sieve of our own experiences and beliefs. The more congruent the facts and stories we share with people are with their belief systems, the more easily they accept them and stack them onto what we already believe.

Consequently, in an organizational change people might know that the facts leadership provides around the reason for the Hoshin are true—the insurance reimbursement has changed, the technology we use has been leapfrogged by a competitor, regulators are demanding an unprecedented degree of accuracy in our design—but some may have strong beliefs that formed when they were laid off at their last company after a major change in strategic direction. Leaders can provide those people with reliable facts and some good stories of how this situation is the same, but also different, and re-form those existing beliefs about organizational changes. The effort may allow the person to filter the change message through a more realistic and current screen.

Finally, the more compelling the storyteller, the more likely we are to digest the bits of fact in what Gottschall calls a "story-consistent" way. We'll interpret the facts the way the most engaging storyteller interprets them. Leaders may want to work toward being the best storytellers. It's human to tell stories, both factual and fictional. What are the implications for our Hoshin? The following list offers some specific ideas related to how leaders can use stories to communicate during the change associated with Hoshin Planning.

Ideas for Leveraging Storytelling during the Hoshin Change

- Catch people's attention with relevant and provocative facts about the market and industry, and invite them to talk. Gazzaniga says, "When aroused, we are driven to explain why."
- Include broader and broader circles of people in presenting the reasons for and facts related to the Hoshin plan at "all-hands" and in other public company settings, to test how well a shared understanding is being built and to make adjustments to the story as needed.
- Fit new information into a schema that already exists, when possible, building a bridge from what is already known to what is new.
- Provide information, or people will make it up. Peoples' ability to interpret and make meaning is only as good as the information they receive. Get the right information in place.
- Make opportunities for people to tell a story that preserves their professional dignity, respects their history and work, and casts them in a meaningful role in the future of the organization.
- Get out in front of the story. Like a piece of sand in an oyster shell, if you introduce real data and rational story bits into the organizational chat mix, people may weave a story pearl around it.

Change and Growth via Hoshin Planning

There is energy and charm and even beauty in the moments of transformation in organizations. As a means for provoking that transformation, Hoshin Planning offers inclusivity, focus, and sometimes profound learning. John Frankovich, Hoshin Planning program manager at NV Energy, is right when he says, "If the organization is truly learning and making changes and executing Hoshin after Hoshin, just think how easy it is to do each successive Hoshin, and people walk out excited and skilled, and then we do it again with more people. Hoshin creates career progression opportunities for people, it builds leaders, it makes us learn together."[13] Well-done Hoshin Planning uses change as a path to growth, learning, personal development, and the building of a community around a shared set of focused work.

Endnotes

1. Holmes, Richard. *Coleridge: Early Visions.* New York: Penguin Group, 1990, p. 93.
2. Bradatan, Costica. "The Wisdom of the Exile." *The New York Times.* August 17, 2014.
3. Deutsch, Karl W. *The Nerves of Government: Models of Political Communication and Control.* New York: The Free Press, 1966.
4. Schein, Edgar H. *The Corporate Culture Survival Guide.* San Francisco: Jossey-Bass, 2009.
5. Coutu, Diane L. *The HBR Interview: Edgar Schein. The Anxiety of Learning.* Boston: Harvard Business School Press, 2002.
6. Coutu, Diane L. Ibid.
7. Schwarz, Roger. *The Skilled Facilitator.* San Francisco, California: Jossey-Bass Inc., Publishers, 1994. Chapter 4.
8. O'Neill, Julia. Merck VAPTO, June 16, 2015, interview.
9. Cowley, Mike and Domb, Ellen. *Beyond Strategic Vision: Effective Corporate Action with Hoshin Planning.* New York: Routledge, Taylor & Francis Group, 2011, p. 6.
10. Miller, George A. "The Magical Number Seven, Plus or Minus Two: Some Limits on Our Capacity for Processing Information." *The Psychological Review,* 63: 81–97, 1956.
11. Gottschall, Jonathan. *The Storytelling Animal: How Stories Make Us Human.* New York: Houghton Mifflin Harcourt Publishing Company, 2012.
12. Gazzaniga, Michael. *Human: The Science Behind What Makes Us Unique.* New York: Ecco—HarperCollins Publishers, 2008.
13. Frankovich, John. NV Energy, June 20, 2014, interview.

Article 7

Hoshin and Acquisition

Contents

> Throughout its history, M&A has been the major game played on Wall Street: the grand chess game that dominates all others. The effects and by-products have had far-reaching consequences—more than most commentators or financiers are publicly willing to admit.[1]
>
> **Charles R. Geisst**
> *Deals of the Century*

The worldwide value exchanged via mergers and acquisitions (M&A) during the first quarter of 2014 was $756 billion.[2] Among those companies acquiring or being acquired, there are those practicing Hoshin Planning or Hoshin-like planning. The secrecy around most of these strategic moves makes a controlled experiment around organizational development and planning methods difficult, but available cases and aggregated survey data support the idea that M&A impacts an organization's planning process in a variety of ways, some very positive, some just disruptive. Whether acquired via private equity, through the investment of company capital, or as part of a subtler practice in big company reorganizations of "acquiring" internal business units, the impact on the Hoshin Planning process and the people leading it can be great.

The ability to incorporate the gains from Hoshin is influenced by many things, such as how far along the organization is in its implementation, whether the leaders who championed the Hoshin stay with the organization, and whether or not a new CEO is committed to the discipline, focus, and participative principles of Hoshin. To help organizations maximize the value from an investment in Hoshin Planning, this article discusses six tactics to consider, listed below, if the Hoshin-related work completed prior to the acquisition is high quality and still relevant to the new combined company's mission and priorities.

Hoshin Planning is built on the foundation of quality improvement principles. An M&A event can trigger an organization's recommitment to those core principles, such as understanding the impact of changes on a system, measuring to understand, really engaging people in meeting the organization's objectives, standardizing improvements and sustaining the gain, managing with a Plan–Do–Check–Act (PDCA) (see Appendix for PDCA) way of thinking, and investing process improvement dollars in strategic directions.

While Hoshin practicing organizations initiated or participated in M&A events in the 1960s and 70s, the method may not have been put under as much pressure from these actions as it is today. In an acquisition economy, how does Hoshin have to evolve to work?

Legal issues from securities law, and competitive sensitivities, that attend M&A events mean that very little information will probably be shared with the people in the organization moving the Hoshin plan through until some time after the transaction closes.

Still, Hoshin planners *can* take some proactive steps to maximize the value of the Hoshin work that is carried forward. This article presents some suggestions. It's important to know your own environment and the conditions of the particular M&A event you're in the midst of when your Hoshin Planning work is interrupted.

- Establish a plan for the Hoshin during the integration planning.
- Maintain Hoshin visibility through measurement.
- Prioritize communication.
- Conduct a review.
- Redirect and adjust resources proactively.
- Consider the human side.

Establish a Plan for the Hoshin during the Transaction and Integration Periods

The work to complete an acquisition transaction is demanding for the leadership levels in an organization, and the development of specific plans to integrate the organization's current Hoshin work can be easy to overlook or postpone. A tendency to underestimate the importance of actively planning a good Hoshin's future is consistent with some organizations' delay of the work to integrate the organizations until the transaction is completed. Yet, having a "well-executed integration plan" is considered the most important ingredient for the success of a deal, ahead of doing "effective due diligence," the existence of "positive economic conditions," and even "correct valuation" of the company being bought, according to the largest percentage of respondents (38%) to KPMG's *2014 M&A Outlook Survey*.[3]

Well-developed Hoshin work, because of its exacting focus on choosing the right enterprise-wide breakthrough improvements for enabling both near- and long-term success, is of the highest and most relevant value to an organization. Hoshin assets ought to be considered as soon as legal and proprietary information sharing limits are eased. Left to chance, those assets and the investment the company has made may be undervalued or lost.

To be clear, most people in the organization will not be in a position to decide the disposition of partially implemented Hoshin objectives. The involved CEOs and a very few others will be working out what will stay and what will go. There can be extreme limits to what information can be shared before and during the pre-transaction and closing periods, and the stakes can be high for moving information around inappropriately. This is particularly so if the companies are competitive with each other. All that said, it makes good business sense to explicitly include a discussion of what will be done with the deliverables of Hoshin Planning, as soon as it is practicable and legal.

In some cases, newly acquired companies are incorporated as mostly independent entities, and they experience a minimal kind of integration, where finances and assets are combined but processes and technology are left mostly intact. In those situations, the exchange of practices and methods may work in both directions, where Hoshin Planning skills and deliverables in the acquired company can move over and add value to the acquiring organization. In other cases, the acquired company is fully consumed, with its products, systems, technology, and so forth sorted through and selected for continuation in an a la carte fashion. What isn't selected is likely discarded. If the acquiring organization is not the Hoshin practitioner, the combined company may see an end to the Hoshin Planning method and its by-products.

During integration discussions, leaders might pay particular attention to what new resources, policies, and processes from the acquiring organization can be used to continue the Hoshin work. What newly available information or methods might help make maximum use of the work done so far and save time in rebuilding Hoshin deliverables later? In addition, if leaders have some idea of how a deal will unfold, the organization can use that information to actively sunset the parts of the Hoshin that will no longer serve the mission, accelerate the completion and incorporation of the parts that will strengthen the new organization, and communicate very clearly to people how things are expected to go down.

Questions for Transaction Period

- What new information or methods arrive with the acquiring organization that can be used to boost the completion of the current Hoshin Planning work?
- What policies in the acquiring organization will impact how the Hoshin is completed?
- What parts of the Hoshin work are now redundant, because the other organization has already done it?
- What parts of the Hoshin plan should be sunsetted, as a result of a shift in mission and priorities?

Clarity on what the landscape will look like after an acquisition is not likely to be available beforehand, but a review and revision of strategies can be conducted well within the first several weeks after the transaction. It can be hard to make the case for adding to the already demanding agendas of mid-stream integration meetings, but the work to weave the Hoshin assets in can save time and rework and possibly motivate good people to stay.

How Hoshin Deliverables May Move into the New Combined Organization

A Hoshin represents a breakthrough in thought or design. Those who didn't help create it may not easily understand it. And because time is often short for moving assets around after an acquisition, some uninformed cutting may come into play. Integration leaders may need to help both organizations see where the Hoshin deliverables can fit and how to value them.

If the Hoshin work has gotten far enough along, it seems to develop its own trail and trajectory in the organization. "People still ask me to show them the Voice of the Customer (VOC) process we used to inform the direction of our Hoshin strategy,"[4] says Julia O'Neill, Hoshin Planning program manager for Merck's Vaccine Product Technical Operations (VAPTO) a year after VAPTO was reorganized and the work on this particular Hoshin discontinued. Methods and concepts delivered by the Hoshin plans move around an organization and find the people who need them. And when the Hoshin objectives are grown from a thorough and honest assessment of the environment and organization and a careful analysis and prioritization of strategies, what results is typically on target for what the organization will need in its future. Unless the actual mission of the organization changes on being acquired, the products of the Hoshin will probably still be good contributions to future success.

At NV Energy, a cornerstone of their *Workforce Redesign* Hoshin was a simplification in labor classifications at their plants. This was a challenging move in a union environment, but it was viewed among leaders as an essential piece of the workforce changes. The leadership team learned new management models, grew to understand each other's priorities and leadership styles, and sustained the demanding work of building consensus around decisions. The Hoshin work done prior to the acquisition transaction paid off. John Frankovich, project director for NV Energy's Generation

Resource Development and the Hoshin Planning program director, explains the longevity of that piece of the Hoshin plan: "The Workforce Redesign had many pieces, or strategies, in motion. The readiness of the job classification change was there at the time of the acquisition, and the training programs to support the change were ready also. The change still made strategic sense, post acquisition. The pieces were ready, and there was a high quality implementation plan ready."[5] It seems to be that if what the Hoshin produces is well targeted and well planned, its benefits should continue after acquisition.

The Hoshin Process and Philosophy after Merger or Acquisition

During the planning that preceded and led to their joint venture with HCA, Methodist Healthcare had deployed such a quality Catchball (see Appendix for an explanation of Catchball) process that it was published and used as a benchmark by others.[6] Methodist's senior vice president, Geoff Crabtree, says, "After the joint venture with HCA, it became more difficult to use the tools of Hoshin Planning, like Catchball, because the organization had grown so rapidly. For example, playing Catchball across eight hospitals, rather than one, has been difficult, simply because of a larger number of employees."[7]

Still, since the joint venture, there has been a concerted effort to sustain the spirit of Hoshin through building all employees' understanding of the expectations and metrics that bring plan implementation activities alive at the department level. Responsibility matrices that plainly indicate outcome metrics play an important role in the newer form of Catchball activity.

Hoshin program leadership and teams may need some time to transition from a Hoshin way of doing things. Merck's Julia O'Neill reflects on the value of sustaining the Hoshin process for a little time after such a big change in the organization, "In the early days of the reorganization, it was hard to give up the Hoshin process. Continuing for a few months helped people get a sense of closure."[8] Hoshin planners manage to take the way of thinking, if not the formal process of Hoshin, with them after a major organizational change like acquisition.

Use Measureable Results to Maintain Hoshin Visibility

As a way to educate a new partner organization on the work accomplished through Hoshin Planning, executives can platform and discuss the metrics used to monitor the Hoshin work prior to the transaction. Hard data are meaningful currency in acquisition discussions and should not be overlooked as a means for building credibility and expressing the value of the Hoshin work.

Let the data speak for themselves. "Safety and energy supply reliability performance in 2014 was the best in the history of the company. The acquisition drove a sense of urgency in our need to come together and get things done. The fact is, eight years of organizational process improvement and Hoshin Planning helped us adapt quickly and effectively to the changes. We were prepared,"[9] says Rich Winters, process improvement manager for NV Energy.

Even if an organization does not recommit to formal Hoshin Planning after a merger or acquisition, the use of Hoshin-like rigorous measurement and reporting is sometimes used to achieve a similar alignment around shared objectives. Geoff Crabtree says, "True alignment within a large organization comes from the implementation of system-wide goals. If everybody is working on a goal platform that is similar, that's how you light the shared flame."[10] Crabtree acknowledges that the longer term planning he enjoyed as part of the Hoshin Planning process before the joint venture with HCA is now harder to focus on in a much larger and constantly changing environment that is more aimed at daily operations. He believes, though, that deploying meaningful strategic metrics, informed by the underlying principles of Hoshin, is resulting in a higher level of service excellence than ever before. Methodist Healthcare won the 2014 Texas Award for Performance Excellence, the state-based precursor to the Malcolm Baldrige National Quality Award.

The quiet private equity giant Danaher Corporation gives managers just a few months to create a clear and measureable strategy and start reporting on it. Their Danaher Business Systems (DBS) seems to equal policy deployment (aka Hoshin Planning), plus balanced scorecard-like measuring and reporting, plus Lean process management. Former CEO Larry Culp expected the entire fleet of companies across its portfolio to apply this management system. So if you weren't doing something like Hoshin Planning before Danaher acquired you, you're definitely doing it once you become part of its portfolio. And something works about it. Among its private equity peers, Berkshire Hathaway and GE, Danaher Corporation jockeys annually for first position in highest return to shareholders.[11]

It's good to feature Hoshin-related metrics in the merger and acquisition dialogue, but take the time to translate them into the language of the new company partner. "Don't make the 20% of the Key Performance

Indicators (KPIs) related to the Hoshin too mysterious if you're readying to sell. Stick to the standards when you want others to understand them and see you as attractive,"[12] says John Henderson, quality director and Hoshin program manager at Gerber Scientific, Inc., a company in the Vector Capital portfolio. Henderson refers to the balance between tracking the business fundamentals that show the current performance of the organization (80% of the metrics we track) and the Hoshin or strategic goals (20%). Those 20% have to be digestible by leaders in both organizations, if the value of the Hoshin process and its deliverables can be assessed properly.

Hoshin Valuation

Can a dollar value be put on Hoshin work? A power plant or a physician practice or a proprietary manufacturing method can be valued as part of an acquisition. Dollars saved from implementing a Hoshin can be included on a financial report somewhere. Even an award is an asset of sorts, something that adds market goodwill to an acquired organization's portfolio. But what is the value of a well-executed Hoshin plan? How about a partially completed Hoshin? As part of a staged long-term plan that is building a foundation for future measurable business results, how can the full value of a particular year's Hoshin be assessed? How do we show the measurable value of an organization's capacity for systematic forward planning?

These questions can be challenging to answer. It's worth asking them anyway, as the discussion that evolves around them can help inform leaders' decisions about whether and how to position the Hoshin work for inclusion in the priorities of the new combined organization. The list below summarizes a few questions to help open the dialogue on how to consider the value of Hoshin.

- What costly processes or practices have we eliminated as a result of the Hoshin work?
- What new revenue have we gained or enabled for the future?
- What building block has been put in place by Hoshin that will provide sustainable growth for the organization in the next three to five years?
- Have we set up a system that, as we grow won't require as many full-time employees (FTEs)? Have we improved our response times or dropped our rework costs?

- What new systems have been installed or are being installed that will take the organization to its next level of performance?
- What measurable gains have been made in employee development and management (e.g., new skill sets, new staffing models) as a result of the Hoshin work?

Prioritize Communication

Inclusiveness is a Hoshin Planning hallmark, and leaders can ease the transition into a merger or acquisition by getting out in front with a message. Delays in providing information produce anxiety, and Hoshin can quickly be seen as too much trouble or too exotic to its new partner organization. Once knowledge that a transaction is going on is generally known, it's time to get moving with delivering what information is ok to deliver. The communication strategies in the list that follows are for leaders who are shepherding Hoshin through at the time of an M&A event. These activities should be included in the overall integration communication plan, not done as a kind of rogue parallel communication process.

Communication Tactics during Integration

- Start to communicate as soon as is legally and practically possible.
- Leverage the heightened visibility and skills of Hoshin leaders.
- Establish one Hoshin-like message.
- Build an ongoing set of Hoshin communication activities.
- Invite a cross-company team to decide how to handle the Hoshin.
- Be clear about what is still open for discussion and what is not.

Leverage the Heightened Visibility and Skills of Hoshin Leaders

When there is a call for someone to present a key operational area as part of the "getting acquainted" period, leaders might consider choosing someone who has been a Hoshin Planning leader. The people involved in Hoshin are often very knowledgeable about the reasons for the direction the organization is taking, up-to-date on innovations in the industry, and skilled in the use of group processing methods and principles. They'll also be better positioned to articulate how the Hoshin work contributes to the priorities of the newly combined organization.

One of the signature features of a Hoshin is that it requires the involvement of most parts of the organization to achieve it. If this is so, then there are many eyes watching and ears listening for how the demanding and satisfying work previously given to the Hoshin will be handled. Both in-place and newly arriving leaders gain in credibility by showing they can anticipate some of the effects of the changes in ownership on the Hoshin Planning process. Hoshin practitioners are particularly well prepared to help in this. John Frankovich describes what he learned through being the Hoshin program manager at NV Energy like this, "There are skill sets I built in the Hoshin process that are helping me now: doing things without ruffling feathers, change management, anticipating what people may try to protect, explaining the perspective of other teams to each other, being an outside perspective. I didn't have that anticipatory viewpoint before Hoshin."[13] Interestingly, these skills are also those of a good integration leader. An M&A event can provide an intensive opportunity to apply them. It makes sense to use Hoshin leaders in the acquisition dialogue. Their skills are good, and they can help two organizations transition toward working well as a combined entity.

Establish One Hoshin-Like Message

Hoshin Planning invites leaders to choose one clear priority for each annual cycle of strategic planning and to communicate that priority throughout the organization as a guiding focus for making decisions and assigning resources. The habit of using a disciplined approach to make this strategic choice for the organization is a good one to maintain, even after the organization is acquired, and regardless of whether that focused style of planning continues to be called Hoshin or not.

Leaders must not only help define a path for the highly focused Hoshin they've begun, but also help identify what the new Hoshin-like priority will be, if it's expected to change. There's value to locking in as soon as possible to that single priority focus that will drive choices as the two organizations integrate. "If you honestly look at the most important thing for your business, in a disciplined way. If you're super focused on the top thing that will make a difference, it's just going to make you better,"[14] offers Kevin Geraghty, vice president for energy supply at NV Energy and the leader who introduced Hoshin Planning into the organization. Getting clear on a common objective and being inclusive are parts of good leadership, but there is more. Strong leaders exercise their voices to embed that one focused objective into decisions that are made in every part of the organization.

Once the focus is regained after an M&A event, it's smart for leaders to create a shared script for how to express the top priority everyone must align around and to use that script for all key communication opportunities. Such scripts include items such as those listed below:

- The current focused priority for the organization.
- Specific guiding principles by which decisions will be made.
- Activities/products/services that will be retired or discontinued.
- How the deliverables from the Hoshin work will be used and integrated into current work.

Consistent delivery of the message is essential. As the organizations merge, confusion and anxiety can already be high. The work new combined leadership teams do to establish a common message is worth every bit of meeting time it takes. "Many more people were involved in implementing the integration than were involved in completing the acquisition transaction. There had to be a common message that could be conveyed by *any* executive guiding *any* workgroup that touched the implementation and planning for the post-acquisition integration to assure the right things were prioritized throughout,"[15] says Elizabeth Liebow, senior vice president for clinical planning at The Dana-Farber Cancer Institute.

Build Some Hoshin into the Integration Communication Plan

To get the dialogue out into the organization, try a structured approach to asking questions. Leaders can hold small group meetings on each question, or have the questions sent out via an employee survey, with a follow-up set of meetings around the areas that surface as major concerns or opportunities. The following list offers some ideas for questions to ask people during acquisition time:

Questions for Soliciting Employee Dialogue

- What would "Done" look like to you for this Hoshin?
- How do you believe accomplishing this Hoshin will make a difference to our position in the marketplace, keep us ahead of the industry?
- What parts of our Hoshin work may be essential to keep as we become part of a larger organization?
- Do you believe that being acquired may present some advantages or help to the Hoshin work we have already begun? Will being acquired make parts of our Hoshin work more difficult or unnecessary to complete?
- Do you believe that your investment in the changes associated with the Hoshin Planning process will make you personally more valuable to the company, to the industry?

In addition, Hoshin leaders might consider creating mechanisms, such as online chat rooms or open sign-up meetings, for people to organize themselves around the topics that concern them most. And when they do come together to talk, the leaders attending should work from that script they created of the details of a shared message. Delivery of the shared message should be monitored and corrected if anyone trails off message—consistency matters.

In addition, routine bimonthly Hoshin leadership meetings can be repurposed to the work of preparing the Hoshin plan for acquisition. The questions listed under the section entitled "Questions for Transaction Period" can serve as a guide. Acquisition time Hoshin leaders might also consider some traditional ways of helping ensure that new people arriving at the organization receive the Hoshin message and those staying on from before the acquisition receive a refresher. Leaders can tap existing routine communication mechanisms, adding a twist to reflect the current acquisition situation and the Hoshin message. Table 7.1 provides a few ideas for boosting current communication activities to support the transition of the Hoshin.

Given the high degree of uncertainty and intensity of work preceding and immediately following an acquisition, it might look inviting to avoid routine communication practices, where the conflict and tension may be high. Avoidance here is a mistake. Plans that are in progress are sometimes abruptly frozen. The pall of a "wait and see" mood takes over. With care

Table 7.1 Acquisition Time Hoshin Communication Twists

Some Routine Communication Activities	*Twists during Acquisition Time*
Department Staff Meetings	Feature the previous and new Hoshin objectives on the written agendas of meetings, even if only as a soundbite.
	Like the increasingly common practice of opening every meeting with a safety story, state the new Hoshin and the reasons it has been chosen.
All-Hands Meetings	Ask every leader who speaks to state the new priority and the logic behind it whenever she or he presents.
	Invite leaders from both organizations to stand shoulder-to-shoulder and express the new Hoshin together.
Performance Reviews	Include discussion on the person's current role in the Hoshin Planning process and key components he or she will continue to have as a personal Key Performance Indicator (KPI).

taken not to make too many promises or solidify decisions prematurely, it's important to resist an inclination to postpone communication. It's best to get people talking.

Invite a Cross-Company Team to Discuss How to Handle the Hoshin

Leaders might consider actively managing the credibility of the Hoshin work by clearly communicating the benefits with their new leadership peers. It may help to think of this work as a kind of "show-and-tell," in which each organization shows the other its harder-to-see assets. This is a good idea during integration, whether or not there is an active Hoshin, as leaders from the two organizations are getting to know each other.

Be Clear about What Is Still Open for Discussion, and What Is Not

Some of the angst after an acquisition is related to which decisions really are left to make and which have already been determined, but not communicated, by leadership. Leaders can alleviate some of the confusion by removing from the dialogue things that are already decided. Mike Anderson, MD, CEO of Commonwealth Hematology and Oncology (CHO), a group of physician practices acquired by the Dana-Farber Cancer Institute, led his physicians, nurses, and staff through the post-acquisition integration. Once he was involved in the planning process of the larger organization into which his group was integrated, he explains, he had to be clear with the leaders and staff he brought with him about which decisions were still open for discussion and which were not. "There were times when I have had to take people aside or in a meeting and say 'this ship has sailed, we have to let it go.'"[16] Not to make this clear can be disruptive and time consuming.

It's a by-product of successfully including the Hoshin Planning work in transaction and integration discussions at the leadership level that some decisions about the deliverables from the Hoshin will have been made before the post-acquisition integration is fully underway. It's also most common in the flurry of meetings in this time that even sincere efforts to be inclusive will sometimes fail, and people will be unclear about what things they will be invited to discuss.

Conduct a Review

When integrating Hoshin work through an acquisition, it may help to remember that Hoshin Planning is a meta Plan–Do–Check–Act cycle, performance improvement conducted at the highest level of the organization and renewed annually as part of the strategic planning work. An acquisition can interrupt the usual calendar of this PDCA cycle and throw the Hoshin work into a premature summative review. Hoshin review meetings can be repurposed to deliver clarity to Hoshin workgroups on what parts of the work are going to be sunsetted and what parts will travel forward and need some planning to integrate well into the organization's work after acquisition. Leaders may want to conduct such a review in as thorough a manner as if the Hoshin were completed and a new Hoshin about to be chosen.

An example of one of the tools that can be used for a lengthier summative review is shown in Figure 7.1. In this example from NV Energy, leaders considered what should be kept, started, or stopped in their Hoshin work at that time.

Doing an ad hoc formal review can also help reboot the thinking about the Hoshin, integrating new information from the acquisition into the formation of the strategies. Through review discussions, new strategies for

Keep	Start	Stop
• Communicate • Messaging to the fleet • Dividing strategies horizontally across the fleet • Leading the Hoshin as a leadership team • Maintain Hoshin scorecard (KPIs) • Face-to-face meetings at launch	• Solicit right change agents to lead strategies—passion—start scouting process. • Develop a visual monitor progress. • Use Continuous Performance Improvement. • Document roles and responsibilities (related to the Hoshin) at the beginning. • Assign an overall sponsor for the Hoshin. • Cross-fleet, but smaller working committee for deployment. • Better local messaging to integrate with the fleet messaging. • Improve two-way communication from plant level. • Provide more face time. • Understand the content of the Hoshin well at the director level. • Provide a more formal, structured project plan. • Coordinate up front to manage relationship with consultant as a leadership team. • Integrate early any existing related efforts. • Educate people on changes they can expect. • Consider phased implementation, if it makes sense to the Hoshin. • Provide more frontline representation on the strategy teams. • Deploy planning down another level of detail; reach front line. • Don't overlap new Hoshin with existing one too closely. • Rotate some champions from existing Hoshin onto new one—supplement sustainability committee with new people. • Scope the Hoshin out at front end—take the time. • Include next level managers in education of directors at front end. • Introduce managers to a mode (Hoshin) that's not going to go away—consider having a co-owner of each strategy. • Show where the Hoshin fits in the overall business strategy.	• Stop too large a working committee • Using same resources on every team—represent all seven dwarfs on a team (except Sleepy)

Figure 7.1 Keep–Start–Stop chart.

accomplishing the Hoshin may emerge. A pause to review and reflect can be very productive and save time later.

Redirect Hoshin Resources Proactively

Money and people will begin to move around almost immediately after the transaction. Resources dedicated to Hoshin may be boosted, maintained, or cut. Which of the three outcomes it is may depend somewhat on leaders' ability to direct those resources thoughtfully.

Very likely, financial teams will fan out across an organization after the acquisition/merger, looking for ways to consolidate and cut costs. This inventorying of dollars may involve one or more rounds of cuts within the first year after acquisition or redistribution. A mentality of scarcity may prevail, budgets may be trimmed, and justifications for spending may be set at a higher threshold. Rich Winters, process improvement manager at NV Energy explains, "The initial 'ask' after we were acquired was significant. At first, it didn't feel like a Hoshin-based set of choices, with analysis of how it would impact the longer term Hoshin plan." Winters adds, though, that "we did need to be doing more measurement and reporting on cost and efficiencies." The adjustment was tough, but he says that after the acquisition "managing the normal challenges any organization has feels less complex now than it felt at the beginning of the process."[17]

John Henderson describes having to cut 8% of the operating budget to meet the expected margins after the first acquisition. "Talk about whacks to the budget." Henderson believes the cuts in most cases were needed and healthy, though.[18] Today their Hoshin Planning metrics are integrated with their tracking and reporting of business fundamentals. Cost cutting after acquisitions has created an acute awareness of the need to innovate how manufacturing and service processes are managed.

While the removing of redundancies and waste should always be going on in a process improvement system, it is especially common practice after an acquisition to see an extra round of clean up. After making a series of acquisitions, Greenville Health System is now focused on what vice president for strategic services Malcolm Isley refers to as the management of risk associated with the day-to-day performance of their work. They planned to take $100 million out of their costs beginning in 2015. He describes this work of finding efficiencies and optimizing processes as "less fun" than generating growth via acquisition. Unlike seeking acquisitions, "There's no revenue bullet around this kind of work."[19] The Hoshin objectives that are chosen after the dust from acquisition settles may feel less exciting, more mundane.

However, after the post-acquisition wave of cuts, organizations may experience a focused and abundant redirection of capital as well. Peter Steinbrenner, director of engineering at NV Energy and the architect of a chief deliverable of the pre-acquisition Hoshin, said his group grew 60% in the first year after being acquired, and had taken on 398 capital investment projects, 220 of which were new.[20] In a similar spirit of tapping a fresh new pool of resources, Methodist's Geoff Crabtree says that HCA, because of its size and scope, routinely identifies best-demonstrated practices in the system and provides a skilled process improvement group to assist hospitals in applying those practices through their local Continuous Process Improvement (CPI) programs.[21]

For Hoshin practices and deliverables worth saving, leaders can take some care that in the frenzy of realigning resources some of those resources are aimed in Hoshin directions? Leaders can try to slow the process down just a little, get business performance data related to the Hoshin that can be included in discussions about the disposition of resources, and show the savings or new revenue that can be contributed through continuing work on the Hoshin.

Consider the Human Side

The early dialogue around buying and selling companies typically takes place around financial statements, not shop floors. As a result, finance capitalism can seem like arms-length play, a bit sterile. It's intriguing, then, how the language around acquisitions can sound oddly intimate. Common phrases like "suitors in pursuit," "midwifing a deal," and merger as "shotgun wedding" evoke frantic courtship and family building. Or at times the conversations sound like planning for war: poaching, predation, full ratchet, golden handcuffs, and the takeover, which of course is sometimes hostile.

When the economists and attorneys are present, words of science and gaming are included: alchemy, synergy, dry powder, anti-dilution, conglomeration, stagflation, due diligence, and game rounds, as in a fresh round of financing or a "down" round, a round of financing at a valuation lower than in the preceding round. These words add a kind of legitimacy or sportsmanship, maybe a sense of objectivity.

Perhaps it follows, then, that the dialogue among Hoshin practitioners before, during, and after an acquisition is often laced with emotional energy and with an effort to tell stories that build meaning from what has happened. In a survey of 100 senior executives done by Mergermarket in 2013, a majority of respondents said productivity dropped and resistance to the acquisition increased when the people part of the integration was neglected. Forty-five percent said neglect of the cultural integration had a "negative impact on share price."[22]

Turnover is key, and alienated middle and senior managers are the most likely to leave during and immediately after an acquisition. When this level in the organization goes, it can take with it the implementation level Hoshin Planning expertise that enables the continuation of Hoshin Planning cycles. Winning the emotional buy-in of these people should be a top priority for serious-minded Hoshin practicing leaders.

Leaders can help keep people by leveraging the skills of the human resources team and making sure to discuss things like benefits, pensions, commissions, and the expected new organizational design. Even if these things change and must be refined later in the integration, people need to know the basics of what their job will likely be and who they will likely report to by the time the transaction closes.

The Intention of the Hoshin May Matter

Hoshin Planning can cultivate a powerful pride among the people who invest their professional equity in completing the process. How that pride is understood, valued, and channeled can be informed in part by how clear people are on why the acquisition was executed in the first place.

The Flipping Hoshin

There is a variety of private equity acquisition that prioritizes a speedy increase in shareholder value over everything else. The thinking in this kind of deal runs something like, "We like the looks of you and will buy you. We can add value by improving a thing or two, and then we can sell you for more money and as soon as possible," not unlike the thinking in a real estate "flip." Being on the receiving end of this kind of obvious logic may dent the collective self-esteem of the acquired organization a little. To be attractive to a certain kind of buyer, an organization may have to show considerable room for improvement. When a company should be run better, it can be obvious to investors from the outside. Being the target of such a purchase can feel like being flattered and insulted at the same time.

The concentration of capital is key to completing a successful acquisition. In a flipping environment, growing shareholder value can become the de facto Hoshin. In such a context, Hoshin objectives like these real ones—*Redefine Physician Relations*, *Improve How We Plan and Schedule Work at the Plants*, and *Build a Seamless Network of Service Providers*—may be less compelling to decision makers overseeing a short-term growth strategy in preparation for a flip. Daily process management, the disciplined, ongoing improvement of the processes that deliver services and products to customers, is generally considered an essential go-with for Hoshin Planning. Process improvement work dedicated to long-term sustainable results, though, may hold less charm in a planning process that is aimed at quickly gaining another point or two of shareholder return. As a consequence, the correction of the types of systemic problems or transformational changes often addressed by Hoshin may be postponed.

In addition, the fact that many acquisitions fail to realize publicly stated strategic objectives, but still make a few investors a great deal of money can sour the atmosphere at an acquired organization. Deals done for their own sakes may feel a little hollow as Hoshin objectives.

Expansion or Diversification Hoshin

By contrast, Hoshin Planning can be a foundation that puts an organization in a very strong position to grow its mission and business through a partnership or acquisition. In a 1995 move that was then unprecedented in the healthcare industry, Hoshin-practicing Methodist Hospital–San Antonio entered into a joint venture with Columbia (now Hospital Corporation of America [HCA]), the largest health care delivery company in the United States. Anticipating dramatic changes in how hospitals would be reimbursed for services in a managed care environment, the hospital's visionary CEO John Hornbeak and his leadership team focused on "creating a larger footprint in order to be more viable to insurers."[23] The unprecedented part of this opportunity was that, in contrast to what had already become a prevalent practice in the industry of larger entities buying up smaller ones, Methodist Hospital entered the agreement as an equal; neither organization acquired the other. Methodist Healthcare's Geoff Crabtree says the disciplines around Hoshin

Planning brought their organization to the table as a partner, rather than as an asset to buy. Such a shoulder-to-shoulder deal was then new in HCA's acquisition history. Crabtree recalls, "We used the Hoshin Planning process to get us into the place where we could look at the potential outcomes of a joint venture as a way to enhance our mission."[24]

In another example, Greenville Health System (GHS) initiated a growth strategy in 2002 with its development of a medical school in partnership with the University of South Carolina. The GHS system has a successful history of applying quality improvement, and in the early 2000s had woven some Hoshin Planning methods and tools into its planning systems. From 2007 to 2014, the system grew from $750 million to $1.8 billion, mostly through strategic acquisitions. "We were focused at that time on building the top line," says Malcolm Isley, vice president for strategic services. Participative Hoshin Planning tools such as the Affinity Diagram and Interrelationship Digraph provided support to the executive team in building consensus around the acquisition plan and deploying it in a way that sustained a positive profit margin, even during the recent recession. Isley explains the effect of Hoshin Planning tools simply, "The Hoshin process allowed every voice to come into the room."[25]

Attractor Hoshin

A high-functioning leadership team leans well into the huge changes that happen when organizations buy, sell, or reorganize. One typical outcome of the completion of a successful cycle of Hoshin Planning is that executives and managers understand the organization's identity and market position with much greater clarity. Renewing or revising the organization's mission and vision, defining critical success metrics, understanding the organization's competitive playing field—these pieces of hard work often produce a robust knowledge of the organization's abilities and its marketplace.

The hard work of Hoshin Planning isn't usually outsourced to a big consulting firm, which will usually deploy a cadre of junior consultants across an organization with clipboards and calculators and eventually deliver a glossy bound strategic plan. Those plans may remain somewhat foreign feeling to the executives who then have to lead them and are often not implemented. With Hoshin Planning, while market research, group facilitation, and other services are often hired to provide outside insight into the forming of the strategy, the heavy lifting of analysis and decision making is done by the leaders themselves. Done well, Hoshin Planning requires the personal involvement and investment of every member of the leadership team. The knowledge and insight earned by doing that work stays with the team when the planning is done. As a consequence of the hands-on work, a leadership team's self-awareness and grasp of the environment in which they're planning can be razor sharp. The ability to articulate this self-awareness is attractive, to both external and internal suitors.

A division leadership team, for example, may leverage the in-depth knowledge gained through the Hoshin Planning process to be more attractive within its own company. As an example, Merck's Vaccine Product Technical Operations (VAPTO) group embraced this Hoshin in 2012—*Build an Integrated Technical Network to Deliver "One-Stop-Shop" Solutions.* Implemented during a tumultuous time for Merck that concluded in a 15% workforce reduction across the company, VAPTO's chosen Hoshin compelled this group to inventory, describe, rationally integrate, and prove the success of the services they provided to the customers they served. This clarity made VAPTO attractive to internal departments when resources and people were being shifted around. The group was still reorganized, but its success in defining its services and carving out an identity facilitated the reassignment of the group's highly skilled staff to other areas of the organization. Julia O'Neill summarizes, "We were hit much less than other groups around us. Our Hoshin process gave us an optimistic, forward-looking approach. That put people in a good light when leadership was considering 'who do we need as we move forward?'"[26]

O'Neill, a Six Sigma master black belt, also likes the subtler process benefits that resulted from using the Hoshin process prior to the major organizational changes. "Even though we've been dispersed, we're all carrying with us an ability we didn't have before. That is, when we have to define who we are as a business, we're able to do it. We brought that to our new assignments. The areas people sit in now are feeling the benefit of the Hoshin work we did before."[27]

Hoshin as an Organizational Organism

Hoshin Planning does something profound to companies, something that is independent of other big changes. New ownership, new leadership, reorganization—these practices will continue as popular means for organizations to grow. The changes can instigate huge shifts in how an organization runs. Still, if a group has completed even one full annual cycle of Hoshin Planning, there is a hum that keeps resonating after an acquisition transaction is executed.

There is something intangible in the room at meetings and in those quiet conversations that are the heart of getting things done in organizations. The formal Hoshin process may go the way of the previous name of the company, but the way of thinking like a Hoshin practitioner has a life of its own. It benefits whatever organization the practitioner stays in or moves to next.

Endnotes

1. Geisst, Charles R. *Deals of the Century.* Hoboken, NJ: John Wiley & Sons, Inc., 2004.
2. *Thomson Reuters Mergers & Acquisitions Financial Advisory Review First Quarter 2014,* 1006243 3/14.
3. Nachman, Sherrie. *2014 KPMG M&A Outlook Survey Report,* "M&A Expected to Rebound in 2014," KPMG, LLP, 2013.
4. O'Neill, Julia. Merck VAPTO, June 16, 2015, interview.
5. Frankovich, John. NV Energy, June 20, 2014, interview.
6. Boisvert, Lisa, and Crabtree, Geoffrey. "Southwest Texas Methodist Healthcare." In Mara Melum and Casey Collett (eds.), *Breakthrough Leadership: Achieving Organizational Alignment Through Hoshin Planning* (pp. 187–214). Methuen, MA: American Hospital Association and GOAL/QPC.
7. Crabtree, Geoffrey. Methodist Healthcare, July 2, 2014, interview.
8. O'Neill, Julia. Ibid.
9. Winters, Rich. NV Energy, May 14, 2014, interview.
10. Crabtree, Geoffrey. Ibid.
11. Bharat Anand, David J. Collis, and Sophie Hood. *Danaher Corporation—Harvard Business School Case Study.* Boston: Harvard Business School Press, 2008.
12. Henderson, John. Gerber Scientific, Inc., October 8, 2014, interview.
13. Frankovich, John. Ibid.
14. Geraghty, Kevin. NV Energy, June 6, 2015, interview.
15. Liebow, Elizabeth. Dana-Farber Cancer Institute, June 4, 2015, interview.
16. Anderson, MD, Michael J. Dana-Farber Community Cancer Care, August 28, 2014, interview.
17. Winters, Rich. Ibid.
18. Henderson, John. Ibid.
19. Isley, Malcolm. Greenville Health System, April 25, 2014, interview.
20. Steinbrenner, Peter. NV Energy, July 2, 2014, interview.
21. Crabtree, Geoffrey. Ibid.

22. Mergermarket. "The Missing Chapter: Why Emotional Buy-In Is Critical for Successful M&A" 2013. The Storytellers. http://mergermarketgroup.com/publication/the-missing-chapter-why-emotional-buy-in-is-critical-for-successful-ma.
23. "Largest Provider of Healthcare Services to the Uninsured in South Texas establishes $1.1 Million Dollar Endowment to Honor Its Chief Visionary and Architect." News release, June 27, 2007.
24. Crabtree, Geoffrey. Ibid.
25. Isley, Malcolm. Ibid.
26. O'Neill, Julia. Ibid.
27. O'Neill, Julia. Ibid.

Article 8

Hoshin Hacking: Planning Lessons from Startup Companies

Contents

There are temporary organizations formed to *find* a repeatable and scalable business model and permanent organizations grown to *implement* a repeatable and scalable business model. Let's call the first one a "startup" and the second one "traditional." Traditional companies have an infrastructure that supports a chain of functions combined specially to bring value to specific customers and fulfill the potential of a chosen business model. Such a custom-made value chain can ensure consistent and profitable delivery of a defined product or service, and Hoshin Planning is sometimes used to add or remove big links in that chain as conditions change in an industry or market.

For example, a manufacturer of laser cutters can make a bold strategic move away from assembling its machines and shipping them intact to customers toward on-demand inventory management and assembling on site for each customer. Such a scale of change meets the three characteristics of a Hoshin: requires a fundamental change in the system, takes support from the entire organization to achieve, and is expected to take the organization to its next level of performance. It's not incremental improvement. It does take innovative thinking to get this done, and it's in the innovation where startups and traditional companies practicing Hoshin may have something in common. In addition, managing parts and assembly in a profoundly different way might also mean being able to break down barriers formed over time by the existing way of operating. Startup leaders might think of that breaking down as hacking around a process, even though their size and mission typically mean they don't have the same kinds of internal barriers to break down. In a big traditional company, though, the scale of such a change to infrastructure would probably also call for a systematic and disciplined planning infrastructure to get it done. And startups really don't like that kind of thing.

"To get something done, you have to use a company's apparatus. To use it you have to use it as it was designed. The bigger a company gets the harder it is to break down the apparatus,"[1] says Tim Rowe, CEO of the Cambridge Innovation Center (CIC), host to several hundred startup companies since its formation in 1999.

Startups are lightweight and energetic, moving like hummingbirds, picking up and depositing ideas like pollen, and hacking new ways around what traditional companies usually do to build products and grow their businesses. If maximized, the special features of Hoshin Planning can allow leaders to move temporarily in this manner, too, supplanting heavy infrastructure with more agile approaches when going after a Hoshin objective. Excellent performers and innovators can be freed from their functional areas for the length of the Hoshin project, barriers to sharing data and designs and information can be suspended, and interesting new connections across the organization can be actively nurtured to generate cross-discipline thinking. In these ways, executives might think of Hoshin Planning as an opportunity for thinking a little more like a startup and gaining the efficiencies and innovation that startups enjoy.

But why should big traditional companies care about how startups do things? Nine out of ten startups fail.[2] There is a magic to startups that is worth considering, though. What other genre of organization could tolerate such a failure rate and still stay so inviting? Because Hoshin Planning deals in breakthroughs that involve risk and doing unprecedented things in the traditional organization, maybe the conditions put around Hoshin work should be different than those around the other work of the traditional organization, more like the conditions of a startup.

Startups invite frugality, an important feature of innovation. Frugal work environments may attract smart, innovative people who don't mind working hard and creating more "shoestring" solutions. Startups are about growth, and they're good at going after big prizes. Maybe blending the best features and practices of startups with the discipline and maturity of Hoshin Planning is a winning combination for traditional companies. To talk about how startups and traditional organizations might learn from each other's planning practices, some words might need reconciling.

Vocab Words

Words like "Lean," "agile," and even "planning" have a history in the quality improvement world from which Hoshin Planning was launched. The startup community co-opts and adapts things well, including words. The result is a lingo that can make sharing learning between the startup and traditional worlds more challenging.

Lean Startup ≠ Lean Six Sigma

Startups use "lean" to describe how they get a minimally formed product to market to get feedback, iterate, send a product out again, adjust, send out again, and so on. Here "lean" is about quick cycles of testing a barebones product or service to get feedback early in the development process and usually for little money.

For the quality improvement community, in which Hoshin Planning generally resides, Lean is about the elimination of waste, or *muda* in Japanese. It's similar in spirit to the startup's slimmed down product development process, but the startup's lean is not as organized, documented, or slow. Traditional companies use Lean primarily to streamline their value chain and deliver their existing products and services to the customer in less time, and at lower cost.

Muda is generally considered to include eight wastes that follow the pneumonic, DOWNTIME—Defective production, Overproduction, Waiting, Non-used employee talent, Transportation, Inventory, Motion, and Excessive (over) processing. Startups use lean approaches to cycle product and service drafts through real customers during the design phase and they probably do it with very little of the DOWNTIME types of waste. But they don't use the word the same as traditional companies do. Startup-style "lean" contributes to the rapid cycles of Plan–Do–Check–Act (PDCA) they are capable of, which is discussed later in this chapter.

Scrum ≈ Kaizen

Consistent with the minimalist thinking of Lean is the software development project management mindset referred to as "agile." It involves small, incremental, steady iterations, over the course of a few weeks or months, never as long as a year. "Agile" takes place through the mechanism of scrum, as in a rugby huddle. Scrum meetings are daily discussions to update on progress to a plan and set short-term goals, usually for that day. Communication is often member-to-member, not to or from a leader. Scrum meetings are usually done standing up, usually last less than thirty minutes, and are usually free of slides or packets of paper.

Scrum meetings look something like Kaizen meetings or "Stand-Up" meetings sometimes associated with Hoshin Planning, with a similar objective of resolving shop-floor-level problems in real time. Both Scrum and Kaizen share the paradigm of "inspect and adapt." In terms of planning, startups also use scrum-like meetings to check in and quickly make adjustments to their "strategic" plans. The closest approximation to something in the Hoshin Planning process might be a daily review meeting. The frequency with which startups complete the "check and act" process is much higher than what is found in most traditional companies.

Planning ≈ Planning

Planning happens in startups, albeit it's a minimal or real-time form of it. Like Hoshin, successful startup planning also seems to deploy one major priority at a time, review frequently, and change course efficiently in response to new information. Startups do it faster, though, and they can make changes to the plans more easily.

Also, while Hoshin Planning values the clarity and monitoring made possible through the use of planning documents, startups are not so into the documentation. "Thirty page business plans make investors think you are out of touch, and they don't want you," a startup founder in the Cambridge Innovation Center's co-working space told me recently. Still, startups *are* making plans, but in ways that are different than what we see in traditional companies. They hack the traditional planning process by co-opting some parts of it and breaking new trails around others. Startups take what they want from the lessons many of their founders learned through the earning of an MBA, and quickly toss out what they can't use.

It's more than just an aversion to documentation that is different about how startups plan. How they are organized, the level of involvement and transparency within the companies, a sense of freedom from conformance to usual regulatory and economic and other constraints, the embracing of failure and the companion ability to change fast, a new style of cooperation, and the entrepreneurial spirit—these are some of the things that light the plans of successful startups on fire.

Hoshin Planning Lessons from Startups

Lightning fast cycles of experimentation, not much planning documentation, total transparency of all information within the company—these things might be tough for big companies to pull off. Still, there may be ways Hoshin practitioners at traditional companies can boost their work by experimenting with some of the practices of startups. The following list outlines a few things planning executives might consider pulling from the startup's wheelhouse.

- **organization**—Create a minimalist organization for Hoshin, and lighten the bureaucracy.
- **Mission and transparency**—Be mission obsessed and share as much information as possible.
- **Plan–Do–Check–Act and reactive versus proactive**—Complete more and faster cycles of experimentation, and get better at responding quickly to changes between how the future was imagined and how it really turns out.
- **Freedom**—Temporarily suspend the restrictions of current conventions.
- **Cooperation and co-opting**—Think more broadly about possible partners and suppliers. Repurpose/redesign successful designs or ideas that already exist.
- **Entrepreneurialism**—Encourage the entrepreneurial spirit among those working the Hoshin.

organization

Startups get their work done with "little o" organizations. They are small, nonhierarchical, and friendly to employee turnover. Their work is completed in intimate spaces described as "disclosive" by the trio of philosophers Charles Spinosa, Fernando Flores, and Hubert Dreyfus in their book *Disclosing New Worlds: Entrepreneurship, Democratic Action, and the Cultivation of Solidarity*. In disclosive spaces, people share an "organized set of practices for dealing with oneself, other people, and things that produces a relatively self-contained web of meanings."[3] Startups create disclosive spaces organically and fast and can change and evolve them easier than larger organizations. Their organizational boundaries typically remain fluid. Like lipids around a cell, the walls of the startup organization are porous but still hold some form that sets them apart from the environment in which they sit. A startup culture seems like a club that is both exclusive and open at the same time.

The forming of a focused and high-quality shared thinking space influences hiring, product and service design practices, and the ease with which people can plan and review together. This self-contained community thinking is encouraged, and some use the word "cult" to describe it, though not negatively. Steve Newcomb was the founder of Powerset, which is now Microsoft's Bing search engine. He explains this kind of culture well in a blog post: "By 'cult' what I mean is a group of super high quality people who trust each other and have similar ways of thinking, learning, reacting, problem-solving and working together."[4]

Wouldn't such a phenomenon be great in a traditional organization, where naysayers and legacy thinkers can sometimes deflate the early work of Hoshin and drag down planning teams? But can larger organizations create a temporary startup-like organization within their heavy infrastructure and free it from some of the conventions observed by the rest of the organization?

Comments from people like Harper Reed of Modest, Inc. suggest that there is merit to traditional organizations at least trying. Reed was chief technology officer for the 2012 Barak Obama campaign. In a National Public Radio interview he explained why governments cannot execute good information technology projects. Among other things, a government procurement process is so slow and convoluted that it's hard to attract the best programmers and project managers to take the assignment. The infrastructure kills the project, sends it over budget and off schedule before it even begins.[5] Can traditional companies use Hoshin Planning as an opportunity to flatten a part of the organization and create a protected space for startup-style innovation to drive how the Hoshin plan is formed?

Some specific "little o" startup practices might be translated to bigger company Hoshin Planning work. The following outlines a few things for executives to consider:

- Free people from other responsibilities for a while.
- Remove some hierarchal conventions from around the Hoshin Planning process (e.g., let someone lead a team their boss may be on).
- Let people talk to whomever they want.
- Suspend some HR rules during the Hoshin cycle (e.g., give a hall pass to Hoshin teams on travel, flex time, speaking engagements).
- Support a new language of shared meaning around the Hoshin. Resist laying the current organizational lingo on the Hoshin work.

Big Company Startup Incubator—NOT

Big companies have experimented with building internal startup incubators for some time. Microsoft, Coca-Cola, eBay, PayPal, Qualcomm, and others have used their big company resource to seed innovative startup companies.[6] That's different than what is suggested in this article.

Hoshin Planning is the most intimately connected operational objective the company has, not a side effort to invest the organization's dollars in an interesting new business idea. This chapter is about reproducing some of the *conditions* of a startup, so that we're sure to nurture innovation in the development of Hoshin strategies and plans. It's about integrating startup-like thinking into the strategic planning of the organization where appropriate, not about encapsulating a startup inside a traditional company.

Mission Obsession

For startups there is no hard separation between C-suite brainwork and the daily mechanics of the organization that get product or service to customers. Founders of small companies get their own hands dirty in day to day work, let their thinking and feelings show, and share knowledge with their employees more easily. People in startups talk and test ideas constantly. Founders are open and transparent and generally share financial and other information freely. They are determined to have an impact, often in a way that is consistent with their own personal values. And the successful ones are tireless in pursuing that impact. In fact, "the most important quality in a startup founder is determination. Not intelligence—determination," the startup company sensei Paul Graham wrote in a 2006 post.[7] Entrepreneurs are often among the most intelligent business leaders, too, but that wouldn't be enough to succeed.

Mission is so powerful and transparent in startups that there doesn't seem to be as much need for written and detailed strategic plans to keep the activity of the organization inside the guardrails. They can plan day-to-day because they never do anything that is outside the range of the mission. Tim Rowe of the CIC says, "I have tunnel vision on building community to advance innovation."[8] This kind of exclusive commitment puts up natural barriers to anything that is not relevant to the mission.

For bigger companies, the need to communicate, align, and manage the work on the plan among dozens or hundreds or thousands of people means that even with a clear and compelling mission, detailed plans are needed. In fact, a greater level of detail can allow for more realistically assessing the impact of the work on the organization's resources and the vetting of the plan by more people. However, leaders in traditional companies can learn from the passionate and ever present purpose and transparency startups bring to their daily work. Below are some mission obsession practices.

- Remind people constantly about the impact the organization was formed to have in the world.
- Answer questions about anything candidly and with as much information as possible.
- Executives, let it get personal—explain why the Hoshin matters to you personally.
- Explain the real financial and organizational situation to everyone in the organization.

Plan–Do–Check–Act Cycle and Reactive versus Proactive

Startups differ from bigger traditional companies in another significant way; they move faster and worry less about the future. In general, startups try things and move on from failures quickly. As an example, Rebecca Xiong, co-founder of SocMetrics, a social media marketing firm, describes an iterative cycle of development at her company: "We had a software product that took us over a year to develop based on customer feedback. It failed ultimately because it was too far ahead of the market need. Then we tried a product offering that took us two months to make. It failed, too, due to lack of clear need. Then we made one in a week. That one targeted the right need and took off."[9] The growth of knowledge and expertise in the team from such cycles is powerful. Each time they stop and try something else, they get better.

These adjustments, or even tossing out a strategy altogether, are called "pivots," the conscious changing of a product or service's design or its path through a market. Pivots are consistent with the way a startup thinks about "Lean." The startup pivot looks somewhat like a stoked up version of what a traditional company might call emergent strategy, the practice of noticing and responding to changes in the conditions around which the original plan was formed. All good companies have probably pivoted or at least have been skilled at considering a possible pivot. Startups do it faster.

Still, some changes really might be best accomplished by big organizations and as part of a long-term strategy that is laid out like foundation stones over time. Even for those kinds of plans, quicker cycles of improvement might be practiced at a micro-level, meaning within individual pieces of the plan. Where traditional organizations get stuck in the Plan–Do–Check–Act (PDCA) cycle, which Hoshin Planning is formed around at the enterprise level, startups just blast right through. For them, the walls that block the PDCA cycle so necessary to successful Hoshin implementation either don't form or are just broken up. They try, they fail, and they try again, and keep the cycle spinning (Figure 8.1).

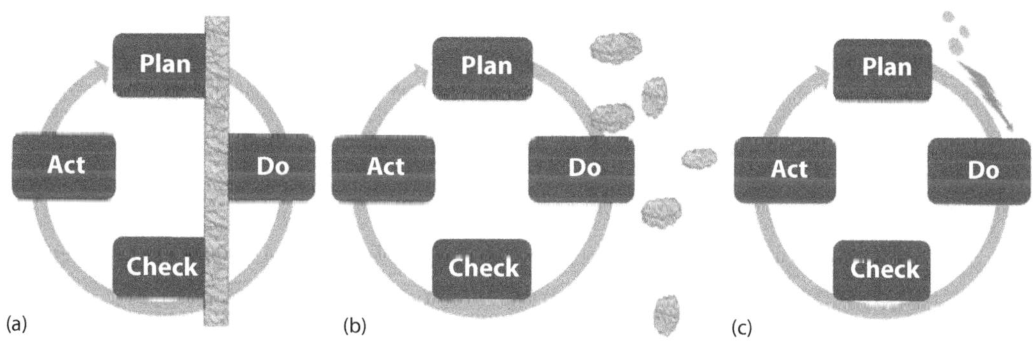

Figure 8.1 (a) PDCA wall, (b) PDCA broken wall, and (c) PDCA fast.

React versus Proact

Concurrent with this spirit of experimentation in startups is a higher degree of comfort with an unknown future state. Our Hoshin plans start with a somewhat detailed picture of the future. We plan backwards from there, outlining specific steps to get from where we are now to the vision we have described.

By contrast, startups value and exercise their skills in something for which traditional companies are routinely chastised: being reactive rather than proactive. The founders of 37signals, Fried and Hansson, in their book of thought-provoking sound bites called *Rework*, emphasize the need for companies building unproven business models to avoid making too many assumptions about the future. "The ability to change course is one of the big advantages of being small. So pay attention to today and worry about later when it gets here. If circumstances change, your decisions can change. Optimize for now and worry about the future later."[10] Startups let the future remain substantially undescribed.

Xiong explains how the difference looks to her: "In a big company, the managers decide on a longer-term business goal, and draw up a step-by-step plan to meet that goal. For us, we aim for an immediate goal, e.g., a product offering that shows clear engagement and revenue potential, then based on what we learned in that process, figure out how best to marshal our resources to grow the business, e.g., to grow that offering or possibly a modified offering in the adjacent market, or to work on something else entirely that we learned is a bigger, better problem to solve."[11] It would be imprudent and unprofitable for executives of traditional companies to think and plan only day-to-day or product-to-product, and to carry their many people along with them on constantly changing courses. But we can create pockets of Hoshin planners that are more agile than the rest of the organization and able to respond to a constantly changing reality more quickly.

To consider building a kind of "Hoshin lab" where speculation and experimentation in strategy, not just technology, can thrive, some questions come to mind. To what extent is the Hoshin breakthrough objective a revolutionary or an unproven business model? To what extent does the traditional company, in this context, need to build and balance an ability to respond quickly with the habit of considering all risks in advance? When a Hoshin is broken down into main areas of work, those workgroups should be encouraged to imagine and innovate. Some experimentation practices include:

- Insisting that the review feature of Hoshin Planning is used faithfully. That's what can bring small company–like agility to big companies.
- Setting up weekly meetings for strategy teams during the first couple of months of the Hoshin. Create Scrum-like review meetings that are short, but direct and productive.
- Insisting that people present their "failures" thoroughly, with explanations and data. Take time at Hoshin review meetings to consider whether those failures might trigger a new way of achieving the Hoshin objective.

Freedom from Conventions, Banality, and Taboos

When Uber and Airbnb started up, regulators and incumbent companies didn't seem to know what to do with them. Who would dare to ignore laws or conventions governing the way commerce is conducted in an industry? How could taxes be levied on a ride in a private car or a stay in a private home? Who would assure the quality of products and services? And before Uber and Airbnb, there was Napster disrupting an entire industry of music production and turning the copyright laws in place at the end of the 1990s upside down. Every industry has its system humming along in harmony, sometimes laced with accepted baksheesh, favoritism, and expensive lobbying professionals. Startups usually consider their business models in a context that is free of conformance to these existing legal or implicit rules. Unencumbered by bureaucracy and limited thinking, startups can consider strategies that often don't even make it to the table in a traditional company.

Keeping Things Strange

There may be some value to protecting a breakthrough in thinking from the constraints of routine patterns of dialogue. Spinosa, Flores, and Dreyfuss write about "banalizers," people who will see a new idea "according to its kind and not its precise content, lumping it in with old ideas, and all the old constraints."[12] Letting an idea stay strange for a while can provide enough space for it to be developed. Maybe we ought not to make it too easy for people to fit our cool idea into their preconceptions too soon. The same may be true with any innovative idea that comes up in the Hoshin Planning process. Protect the "breakthrough" from the banalizers until it develops some.

Similarly, it might be a hazard to breakthrough components of a Hoshin plan to explain the logic behind them too well or too soon. Spinoza et al. hypothesize that if a new thought is "formulated carefully, it will seem obviously true. And this will stand in its way."[13] Naysayers will discount an idea as something that must have already been considered, as it makes so much sense. There is value to maintaining a dialogue around how things looked before this breakthrough idea was formed and keeping the new possible state looking distinct and strange. Startups have an openness and sensitivity to recognizing anomalies, things that don't fit into patterns to which people are already accustomed, and then maximizing the strangeness around the solution they form. It's in the strangeness that breakthrough thinking often happens.

Timing the sharing of an idea can influence how well it takes off, too. Paul Graham in his book *Hackers and Painters* suggests people not to share and defend an idea too soon: "Argue with idiots, and you become an idiot."[14] For Hoshin practitioners, if the Hoshin is heretical to existing practices, maybe let it sizzle for a while in the protected space of the Hoshin teams before engaging the naysayers. If Hoshin is about "breakthrough," then cultivating a safe space for innovation is part of what it means for executives to nurture breakthrough thinking.

Taboo

Things people think and say that are outside accepted practice and that get them in trouble are called taboos. In the thought space created around taboos, there can be some crazy non-mainstream, and winning, ideas. If we agree with Paul Graham, "To do good work, you need a brain that can go anywhere. And you especially need a brain that's in the habit of going where it's not supposed to. Great work tends to grow out of ideas that others have overlooked, and no idea is so overlooked as the one that's unthinkable."[15] Both in the early stages of Hoshin when new system designs are being considered, and all along the way when various means for implementing new designs are being used, there may be merit to encouraging some taboo thinking. What ideas that may previously have been considered verboten can be tossed around as part of the dialogue?

The point is not to break laws, but that it might be possible to think outside existing restrictions when working on a breakthrough for the organization. The following list summarizes some ideas for how traditional organizations practicing Hoshin Planning might actively free some space for innovative thinking in the planning:

- Question existing assumptions about what is and is not acceptable, what can and cannot be done.
- Be heretical; discuss taboos related to your industry and company.
- Question authority, including your own, about what is doable.
- Protect the "out there" idea while it incubates.

Cooperation and Repurposing

Startups are generally great at using social exchange to develop and vet ideas and get funding. They seem to do at least two things that traditional companies might try to do better: staying open to who they can cooperate with, and repurposing others' ideas.

Cooperation

Ironically, the makers of apps that are sometimes accused of desocializing the world by focusing people on their phone screens generally have to interact constantly to advance their ideas. The Internet is a big part of what enables any company to immediately leverage an extremely diverse set of skills and resources outside its organization when building and selling its products and services, instead of building everything in a resource-intensive way internally. Those who do so "hack" design, distribution, or delivery processes that are in place and invent new ones. This way of thinking is consistent with the spirit of Hoshin Planning as a breakthrough strategic planning method and can be allowed into the dialogue of executives and Hoshin workgroups.

In her post "Capitalism, Meet Cooperative Capitalism," ZipCar founder Robin Chase describes the dynamic exchange environment of today's startups. She calls the fluid network of virtual and real partners in which startups exist, Peers Inc and a "transformation of the relationship between companies and people." Loosening the boundaries around a company can let in a diversity of thinking and boost internal capability.

> Peers Inc finds abundance where there once was scarcity. It leverages the ability of individuals and small actors to experiment, adapt, iterate, and evolve. When done well, Peers Inc can create change at a pace, scale, and quality we previously thought impossible. Peers Inc is leading the transition from industrial capitalism to the collaborative economy.[16]

Some observers of startups say that startup founders partner too quickly and break up too slowly. Larger traditional organizations can have the opposite problem, though. Traditional companies are often too slow to break old network rules and partner outside their safety zone and too fast to cut out a supplier or partner who doesn't quickly and obviously fit the apparatus that supports the priorities newly identified by the Hoshin. Below are some creative cooperation ideas.

- Consider accomplishing the value chain in more ways that don't depend only on internal resources and abilities.
- Break through the boundaries of the organization's network and talk with new people.
- Reevaluate legacy relationships that don't contribute to the future Hoshin is evolving toward.

Repurposing

A corollary of cooperation is what Spinosa et al. call "cross appropriation," taking practices or designs from others and repurposing them, as all ideas are considered fair game. "I've been told that I've reinvented the wheel," says Michael Burtov, CEO and founder of GeoOrbital, maker of an electrically powered bike wheel. And he has. But he didn't invent the wheel, the bicycle, or electric-powered locomotion. He puts those things together in a new way that solves a problem. "Nothing is from nothing," adds Burtov. "We're all standing on the shoulders of others. Edison didn't invent the light bulb, he improved on an existing patent and made changes that made the light bulb more practical."[17] About 90% of patent applications are for improvements on existing patents.[18] Startup founders are improving on other people's ideas all the time. Hoshin groups seeking breakthroughs might consider what is already out there that might be tweaked to produce a breakthrough for *their* business.

Entrepreneurialism

Business is personal for entrepreneurs. Most of them are doing what they love. They are attracting people to the impact they want to have. And they communicate that impact in ways that help people understand how their solution will fix a problem. It's more than having a good idea. Plenty of people and companies have intellectual property that doesn't turn into a viable business. It takes, at least, personal engagement, magnetism, and the ability to communicate in the language and at the emotional pitch of their users to build an idea into a successful business.

Personal Engagement

Entrepreneurial personalities are good at recognizing and leveraging disharmonies. Disharmonies are things we overlook, because they don't fit in an obvious way with our other ways of doing things. Entrepreneurs are attracted to and excited by these anomalies. They're intensely engaged with their environment. It's a noble-sounding gift entrepreneurs have, described by Spinosa et al. as opening "new space for human action."[19] They change the way we see things and how we can do things. They don't follow change, they make it happen. Founders are usually people with a natural predisposition toward seeing things differently.

In addition, entrepreneurs seem especially skillful at responding to the present, are sensitive to current conditions, and value having a personal experience of things for themselves. And then they must also have an interest in opening the idea to others. "Startups can not only change the material situation of our lives, but also initiate new styles of behaviors. They make history,"[20] continue Spinosa et al. And they do it by engaging other people on the trip with them.

Magnetism

Entrepreneurs who succeed in planning are able at some point to normalize the anomalous into an organizational environment. Entrepreneurs create interesting disclosive spaces, fill them with interconnected tools and ideas, and engage people in achieving a shared purpose. Those spaces are so specialized that they attract others who are like-minded. Passion and a compelling vision bring people into the startup's space. "Nothing you can really say definitively. But what you can say is that if you don't have a passionate person who sees their world as revolving around success with the new idea, with the fear of god if they fail (unemployment), it usually doesn't work,"[21] says Tim Rowe.

Communication

Telling of experiences or ideas that are very different from typical ways of thinking, explaining why those ideas have become viable, and showing how to relate a new idea to the current situation, these are messages included by entrepreneurs when talking with their teams, investors, and customers. They create sensitivity to anomaly in other people that is similar to the one that led to their insight or innovation in the first place. They don't seem to press their ideas; rather they share their experience and sensitivity through personal stories.

In addition, entrepreneurs are often good at imagining themselves in the position of their users. Paul Graham emphasizes a factor successful hackers have, empathy: "Most makers make things for a human audience. And to engage an audience you have to understand what they need. Nearly all the greatest paintings are paintings of people, for example, because people are what people are interested in."[22]

Perhaps leaders in traditional companies can stimulate this kind of empathy in Hoshin planners. A focused curiosity about how their customers think, what those customers feel strongly about, and what they value offers the promise of building products and solutions that resonate well in an organization's chosen market.

Is There a Hacked Hoshin for Startups?

Let's consider two bits of Hoshin that might add value to the startup founder's limber planning practices: Catchball and data-based decision making. Donny Askin, former CEO of CommercialWare (now part of Oracle), OrderMotion (now part of Netsuite), and currently principal at TechTurn Consulting, a firm serving early and growth stage companies, says finding the right level of planning structure for a startup is tricky. "It's like eating prunes. Is one too few? Is three too many?"[23] offers Askin. Postmortems on all those startup businesses that fail are plentiful and candid. It seems that businesses that win really don't ignore business processes, structure, and planning altogether.

Angel investor and analytics expert Neil Patel advises startup leaders to step back, get a big picture view, and consider process a little more.

Some of the most important components of a startup are those pesky issues of business process, business model, and scalability. Successful entrepreneurs understand that they must work on their business, not in their business. Getting caught up in the minutiae of presentations, phone calls, meetings, and emails can distract the entrepreneur from the heart of the business.[24]

Hoshin Ideas for Startups

- "Catchball" to leverage how suppliers and stakeholders are brought into the planning
- Data-based (not databased) decision making

Bringing Suppliers and Stakeholders into the Planning with "Catchball"

How do startups engage their partners and suppliers in their plan? The Hoshin process "Catchball" helps leaders be clear with people about what type of feedback is wanted and then organizes the input in a way that can be used. Catchball can evolve a general "What do you think of this idea?" into "What do you think of these specific ways of making this breakthrough real, and these ways of measuring whether we achieve it or not?"

Add a Little Documentation

To get feedback on a plan, not just an idea, through Catchball, some minimal amount of documentation is needed. One-page plans are one of the conventions of Hoshin Planning that might fit especially well with startups. The classic Hoshin Planning template includes sound bites of the essential information needed to solicit input from those who will either be responsible for implementing some of the plan, partners in delivering things that will impact the plan, or people who may be recipients of the results of the plan. In big companies, it's essential to continue to populate the one page plans at successive levels of detail. Figure 8.2 shows how the one-pagers cascade into more detail that can be sent out to advisors and around to company employees and then back up to founders.

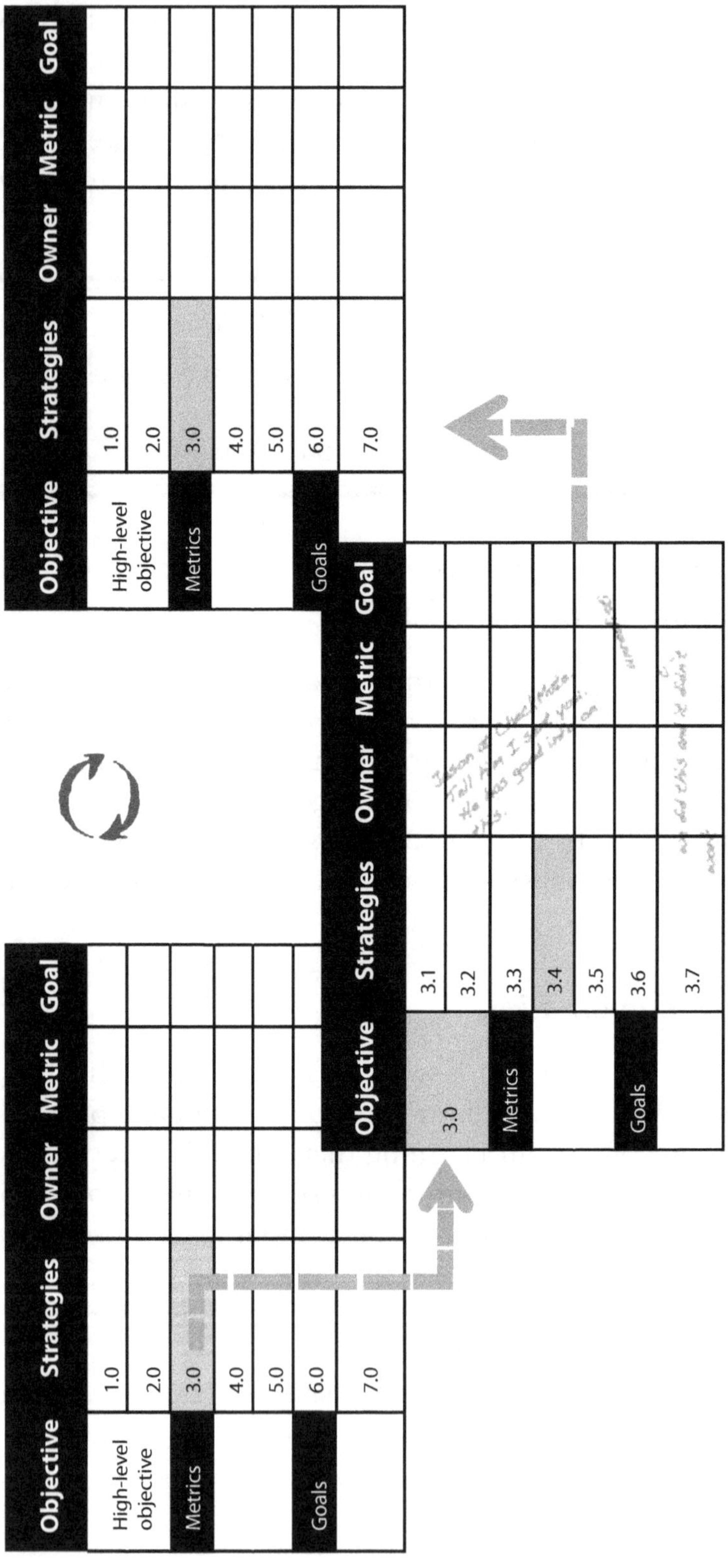

Figure 8.2 Catchball flow.

Elements of the One-Page Hoshin Plan

1. Overall objective
2. Metrics for how success will be measured
3. Main strategies or tactics that will be taken to achieve the objective
4. Who is accountable for each strategy or tactic

In startups, just a couple of levels of plan documentation might be enough, whatever degree of detail builds accountability into the plan. The framework of the form adds organization and focus to the process of soliciting useful input and document very clearly who is accountable for what measures of success.

Data-Based (Not Databased) Decision Making

Startups are good at understanding their metrics at the customer level. Ask about their churn rate, the rate at which their customers drop them, and you'll usually get a confident answer. Or registration-to-purchase conversion rates. Or click through rates. But ask about the types of outcomes metrics used in longer term strategic planning, and the answer may not be as quickly forthcoming. Why should it be? Successful late-stage startups are bought, and founders often are not responsible for sustaining a plan through further growth. Why move the longer-term dials?

> **"You make what you measure."**
>
> Merely measuring something has an uncanny tendency to improve it. If you want to make your user numbers go up, put a big piece of paper on your wall and every day plot the number of users. You'll be delighted when it goes up and disappointed when it goes down. Pretty soon you'll start noticing what makes the number go up, and you'll start to do more of that. Corollary: be careful what you measure.[25]
>
> **Paul Graham on the Y-Combinator blog**
> *February 2009*

Startups may be at risk for being pinged around by measurement, tracking and responding to changes in data in a way that just meddles with a process without really improving it. Any size company has to be clear about what is learned from a measure and how ready it is to act on what it learns. To measure for the sake of measuring wastes time and money. There may be a couple of ways better use of Hoshin-like measurement can support startups: supporting making decisions about the business model, and providing early cues when things are going south.

Decisions about the Business Model

Decisions like "Free, Freemium, or Paid?" may be aided by quick cycles of experimentation where some of the evaluation of success is objectively measured. "Vanity" metrics, like the number of hits to your website make you feel good but may not inform how you should make a decision. In contrast, split tests, such as the A/B test, provide data that informs a decision. If you're a Web-based startup, show half of your site users a new feature, don't show the other half, and see which half's revenue is higher after the test period.

Having clear and measurable criteria for recognizing success might inform a startup when it should pivot or when it should stop making changes and stabilize for a while. There is some merit to having a data-based way of knowing when to pull the plug on the change. Knowing at what point there is a real trend in data can allow for better timing.

Cascading Metrics and Leading Indicators

Is it helpful to link metrics together with a longer term one at the end, to think about what dial needs to move today to make another one move down the road? Does a longish view contribute to the sustainability of a company's business model? Could startups use a little more numerical warning when things are about to tank?

Savvy companies in the startup environment keep an eye on what traditional companies might call interim or leading metrics, those metrics that provide an early signal that something good or bad might be coming. "I track things more than I plan things. I know the workload of every employee and the satisfaction of every employee, because we ask each employee every month with an e-survey. This can give me a heads-up for bigger problems, like a project is failing or that we have a customer service problem in one area of our business,"[26] says Tim Rowe. Aiming some of the measurement energy toward slightly longer term dials may make a startup appear to an investor a little more sustainable or viable, that what they're doing is seen as a platform for continued growth in specific and measureable ways.

Hoshin Planning as a Flexible Planning Practice

To be clear, by Hoshin hacking I'm not recommending skipping anything in the Hoshin process, which is already clean and efficient when it's practiced right. The hacking here mostly refers to what might be done to the organization and culture surrounding the Hoshin process and the willingness to chip away at some of the less functional parts of the organizational environment in service to applying Hoshin Planning principles and tools *more* fully, not less. The spirit, orientation, and focus of startup companies might be co-opted to some extent for Hoshin practicing traditional companies to add fuel to their Hoshin Planning work.

Endnotes

1. Rowe, Tim. Cambridge Innovation Center, May 14, 2015, interview.
2. Patel, Neil. "90% of Startups Fail: Here's What You Need to Know about the 10%." *Forbes*, January 16, 2015.
3. Spinosa, Charles, Flores, Fernando, and Dreyfus, Hubert L. *Disclosing New Worlds: Entrepreneurship, Democratic Action, and the Cultivation of Solidarity*. Cambridge, MA: The MIT Press, 1997.
4. Newcomb, Steve. "Cult Creation," 2010. http://www.BlogNewcomb.com.
5. National Public Radio. November 1, 2014.
6. Miller, Ron. "Coca-Cola Hopes Its Startup Incubator Is The Real Thing." *TechCrunch*, November 10, 2014.
7. Graham, Paul. blog post 2006.
8. Rowe, Tim. Ibid.
9. Xiong, Rebecca. SocMetrics, June 23, 2015, interview.
10. Fried, Jason, and Heinemeier Hansson, David. *Rework*. New York: Crown Business, 2010.
11. Xiong, Rebecca. Ibid.
12. Spinosa, Charles, Flores, Fernando, and Dreyfus, Hubert L. Ibid.
13. Spinosa et al. Ibid.
14. Graham, Paul. *Hackers & Painters: Big Ideas from the Computer Age*. Sebastopol, CA: O'Reilly Media, 2004.
15. Graham, Paul. Ibid.

16. Chase, Robin. *Peers Inc: How People and Platforms Are Inventing the Collaborative Economy and Reinventing Capitalism*. New York: Public Affairs, June 9, 2015, p. 18.
17. Burtov, Mike. GeoOrbital, June 22, 2015, interview.
18. Canadian Intellectual Property Office. http://www.ic.gc.ca/eic/site/cipointernet-internetopic.nsf/eng/h_wr03652.html?Open&wt_src=cipo-patent-main&wt_cxt=learn#part1.8.
19. Spinosa et al. Ibid.
20. Spinosa et al. Ibid.
21. Rowe, Tim. Ibid.
22. Graham, Paul. Ibid.
23. Askin, Donny. TechTurn Consulting, March 4, 2015, interview.
24. Patel, Neil. Ibid.
25. Graham, Paul. "Y-Combinator Blog." February 2009. http://paulgraham.com/13sentences.html.
26. Rowe, Tim. Ibid.

Appendix: Hoshin Tools

Contents

Hoshin Planning is enabled by a linked set of management and planning tools. This special set of tools provides a means for a leadership team to build and review the content of the plan as a group. In many cases the output of one tool is the input of another.

For comprehensive instruction on the construction and analysis of these tools, readers may like to see *The Memory Jogger II: Tools for Continuous Improvement*, *The Memory Jogger Plus*, and *Coach's Guide to The Memory Jogger II*. These three practical job aids provide step-by-step guides and real examples to help Hoshin practitioners leverage the use of the tools.

Each tool section in this Appendix opens with a conceptual graphic of the tool and closes with an example. The examples are based on the work of a real organization and are linked together to show how the tools work in a flow. For clarity and privacy, the content of the planning tables has been added, edited, or deleted, and proprietary information has been removed.

Plan–Do–Check–Act

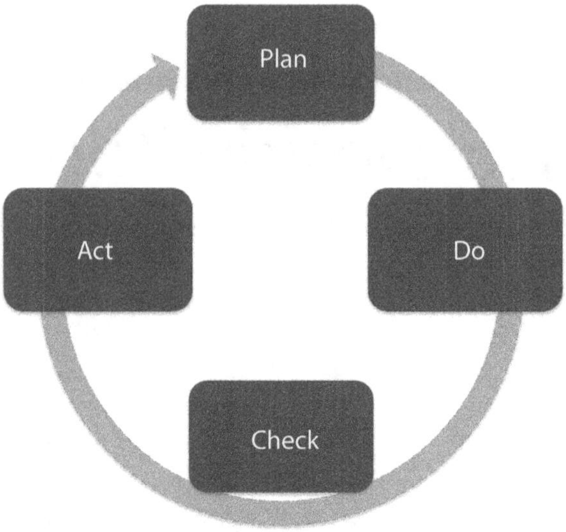

In the practice of science, a researcher studies the available data and information from the system around a problem, forms a hypothesis about its cause, experiments with solutions, and then provides feedback into the system. In the 1930s Walter Shewhart understood that the cycle of thinking formed around scientific research could also be applied to the study of industrial quality control. He called that cycle the Plan–Do–Check–Act cycle (see above) and used it as a foundation for the study of quality improvement in organizations.

In the 1950s, W. Edwards Deming moved the way of thinking represented by PDCA into greater popularity through its use as the foundation of statistical process control. The promise of consistent use of PDCA in resolving manufacturing and service problems was that by identifying and monitoring measures that reflect the health of a process and then identifying the root cause(s) of problems with that process, workers would choose and develop solutions with better specificity and effect, saving time and money and improving quality.

Relevance to Hoshin Planning

Hoshin Planning is process improvement at an enterprise level, an application of PDCA for the whole business. In Hoshin the cycle is often initiated with a "Check," in which a balanced (financial, operational, human) set of data on the organization's current performance is gathered and analyzed as a means for identifying enterprise-wide issues. This check is used to inform the formation of the Hoshin plan.

The PDCA cycle represents a closed loop. It is implied that real improvement cannot be achieved without completing the trip around the loop. The commitment to completing the PDCA cycle is also at the heart of organizational learning, particularly in the "Act" part of the cycle. If the data (Check) from the implementation of our plan are good, perhaps we act by standardizing it across our fleet or training everyone on how to do things the new way. If the data from our implementation are not good, perhaps we act by discontinuing the implementation and conducting another analysis.

With this active review and adjustment, Hoshin plans are kept on track. The Hoshin review table is structured around PDCA, asking those reporting on the plan to identify what was planned, what the actual performance is, the causes of any deviations from the plan, and the countermeasures for correcting the deviations. The review process in Hoshin is essential to closing the PDCA loop.

Affinity Diagram

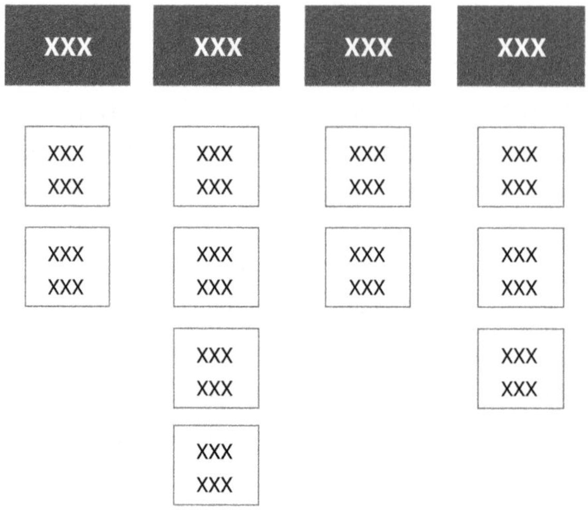

Tool Purpose

The Affinity Diagram is used to organize a large number of ideas generated during a brainstorming session. The tool is likely based on the KJ Method developed by Jiro Kawakita in the 1960s, and it emphasizes every participant writing their own ideas and posting them in a central space where they can be organized by the group together.

The Affinity is an inductive rather than deductive tool. That means that those using the tool don't begin with set categories into which they organize ideas, but rather allow the categories to emerge on their own through a silent sorting process. The initial lack of structure provides an opportunity for leadership teams to think beyond the usual categories of planning and organization. A silent sorting process particular to the Affinity Diagram allows teams to produce a final product that orders the collection of ideas into a rationale set of activities for inclusion in a plan.

Tool Mechanics

Typically the Affinity Diagram is completed using sticky notes on a flat vertical surface, but people can use index cards organized on a desktop, too. The important thing is that everyone's ideas are recorded on the same exact media and with the same type and color pen. These simple conventions help ensure that everyone's ideas are considered equally.

The brainstorming is conducted around a framing question. The crafting of that question is very important, because people will generate answers to that particular question and not another. An example of a framing question for the first Affinity Diagram a leadership team produces in the Hoshin Planning cycle follows.

> It is five years from now. We have met our business and organizational metrics. We have leveraged our strengths, eliminated our weaknesses, taken the best and relevant opportunities, and blocked threats to our success. What does it look like? What did we do?

When the brainstorming is done, participants begin to sort the ideas. The categories that form from the ideas are labeled with what are called "header cards." The header cards are then used in the next tool in this chapter, the Interrelationship Digraph and also the Radar Chart, to determine which categories are drivers of others and what the organization's current performance is in each. Those two tools help a leadership team choose a Hoshin objective.

Uses in the Hoshin Planning Cycle

Brainstorming is one way to identify the vision, strategies, and tactics of a Hoshin plan. The Affinity Diagram is used repeatedly in the Hoshin Planning cycle to process the results of these brainstorming sessions. First, it is used to help a leadership team create the "Vision Affinity," the collection of detailed items that together form a vision for the organization that reaches out three or five or more years.

The tool is applied again to brainstorm the main strategies for achieving the Hoshin, once the Hoshin area is selected. And then it's used again by strategy teams as they brainstorm and organize the specific tactics needed to accomplish the strategy to which that team is assigned. The exact same process is repeated throughout the development of each level of the plan.

Figure A.1 shows a real Affinity Diagram from the power generation company whose work we'll carry through the Appendix as a case study example. Figure A.1 is a Vision Affinity, meaning it was completed at the very start of the Hoshin Planning process in answer to the questions indented above: "What does it look like? What did we do?" This describes a concrete vision of the organization's future.

Considerations

- Both the mechanics and the shared experience of learning are strengthened by building an Affinity Diagram together, with the entire team collocated.
- This tool takes time to do well. Allow about two hours.

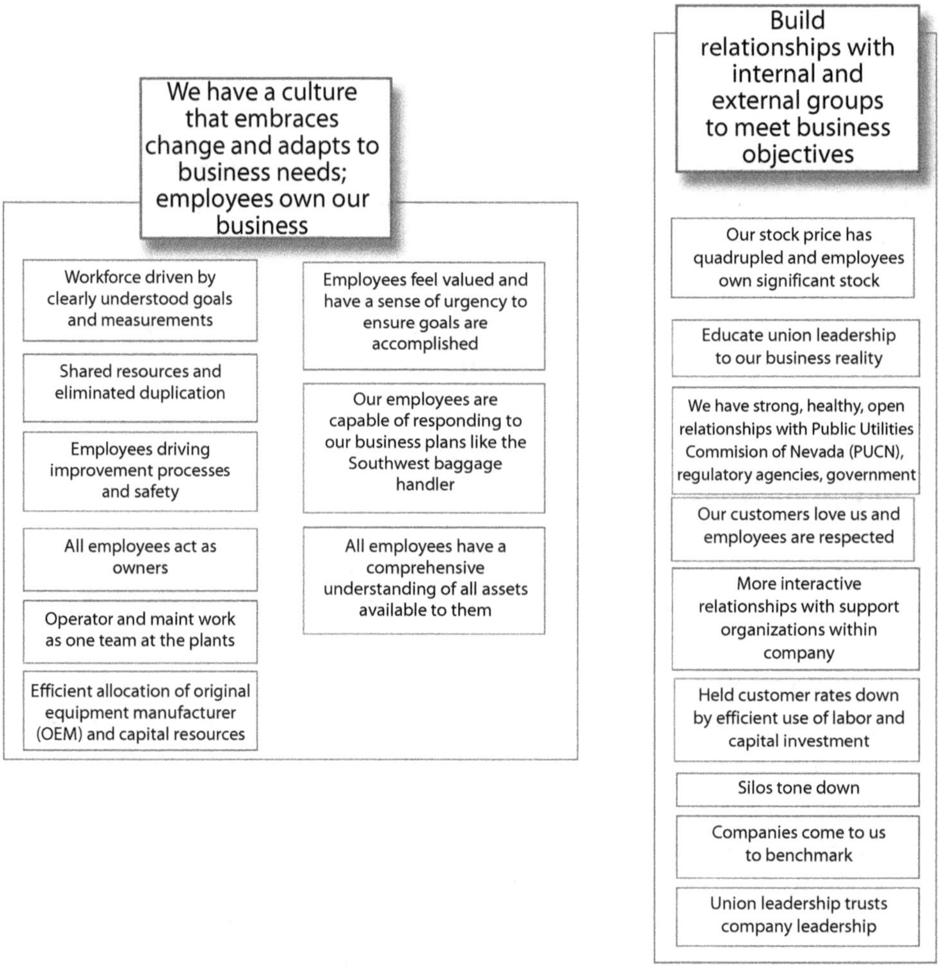

Figure A.1 Vision Affinity Diagram case study. ***(Continued)***

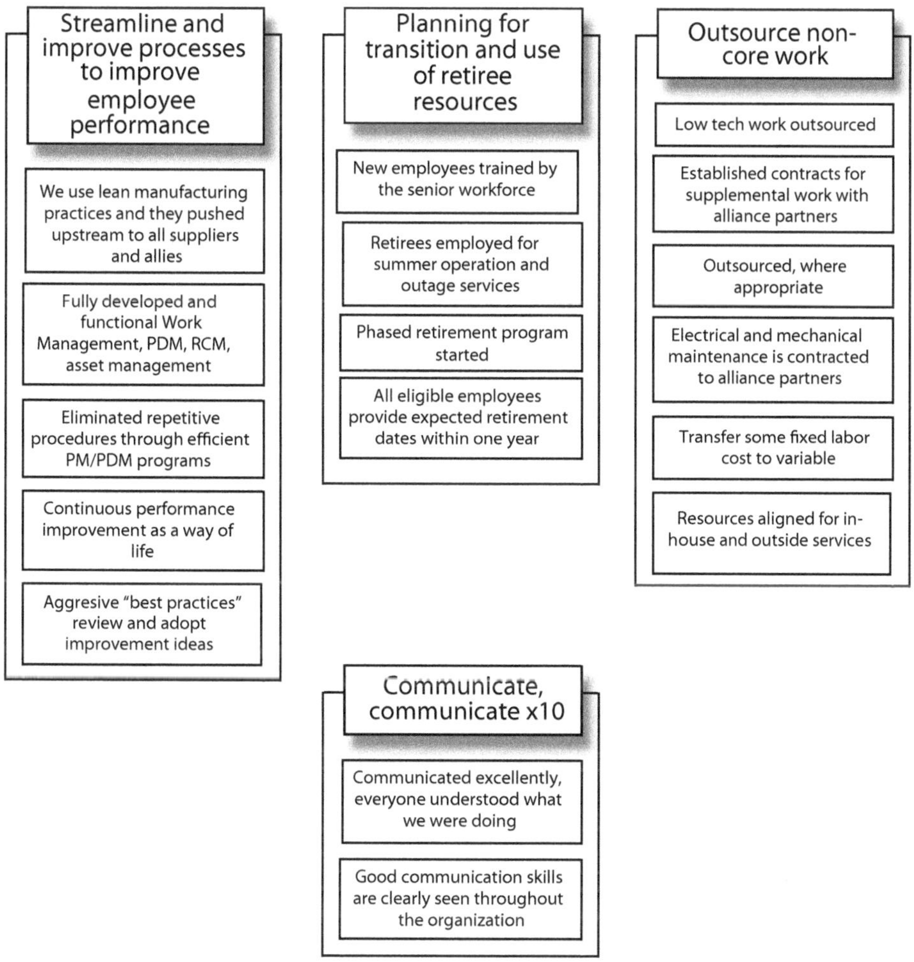

Figure A.1 (Continued) Vision Affinity Diagram case study. *(Continued)*

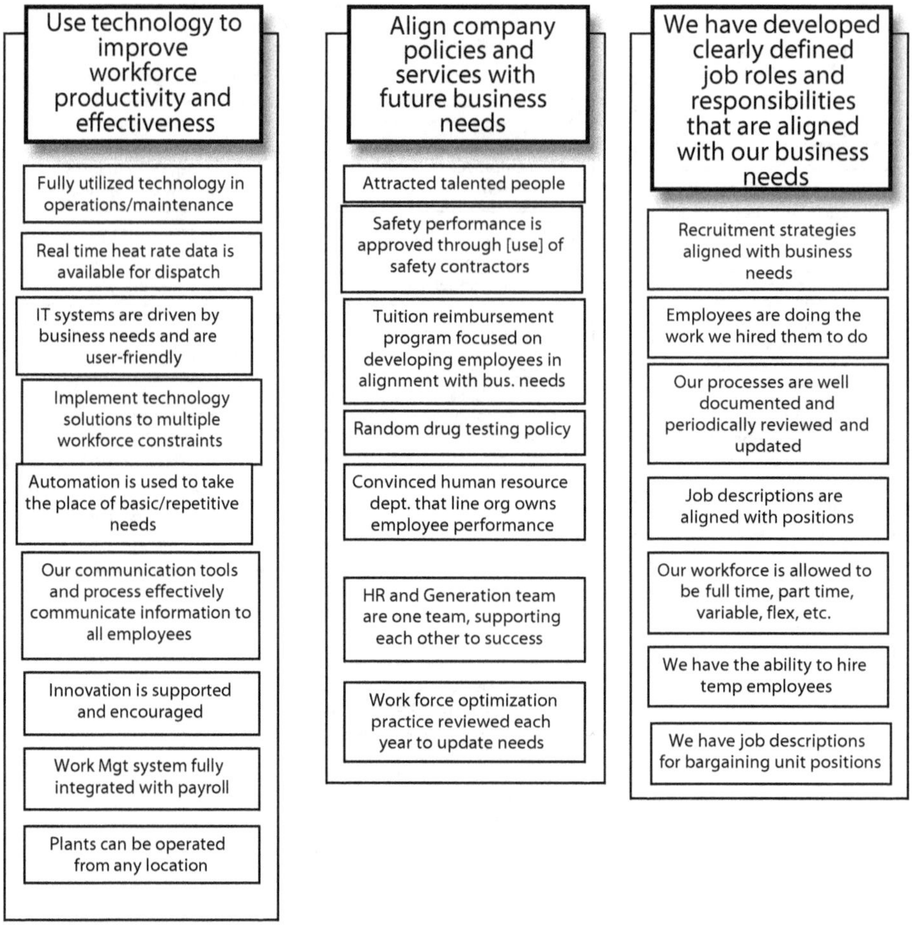

Figure A.1 (Continued) Vision Affinity Diagram case study. ***(Continued)***

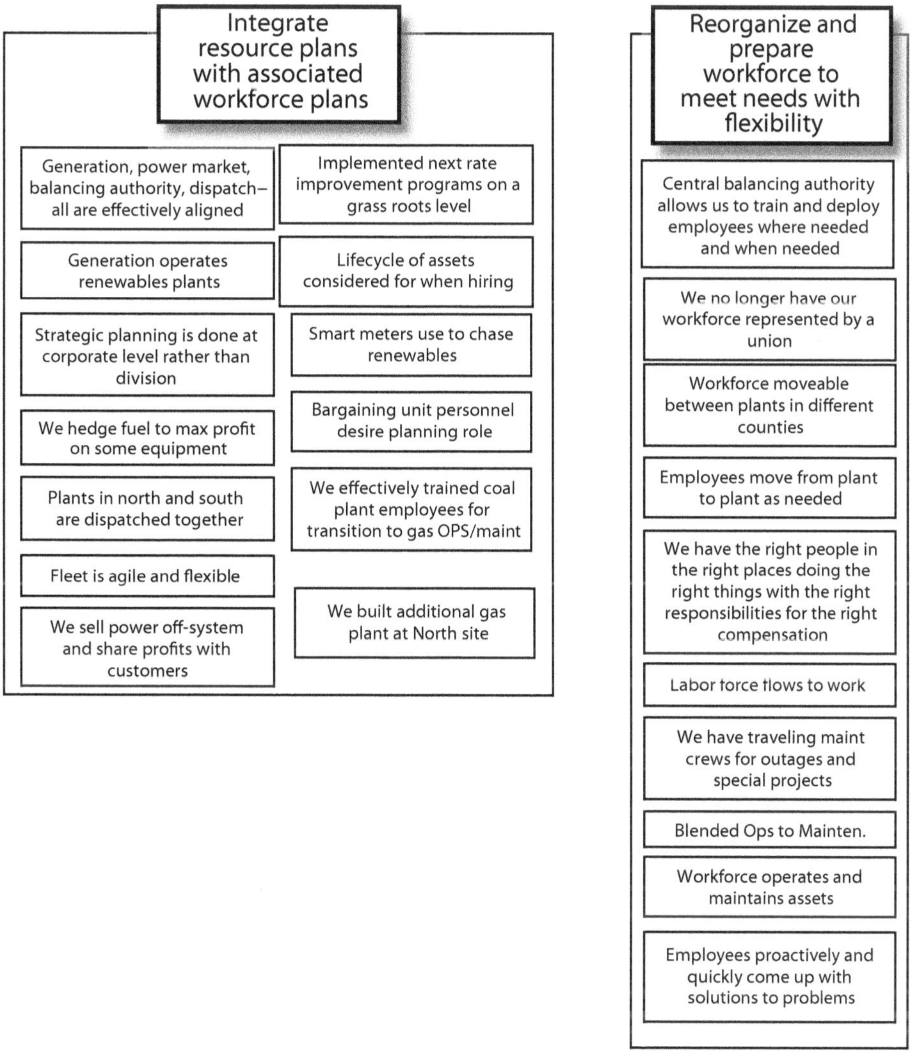

Figure A.1 (Continued) Vision Affinity Diagram case study. (*Continued*)

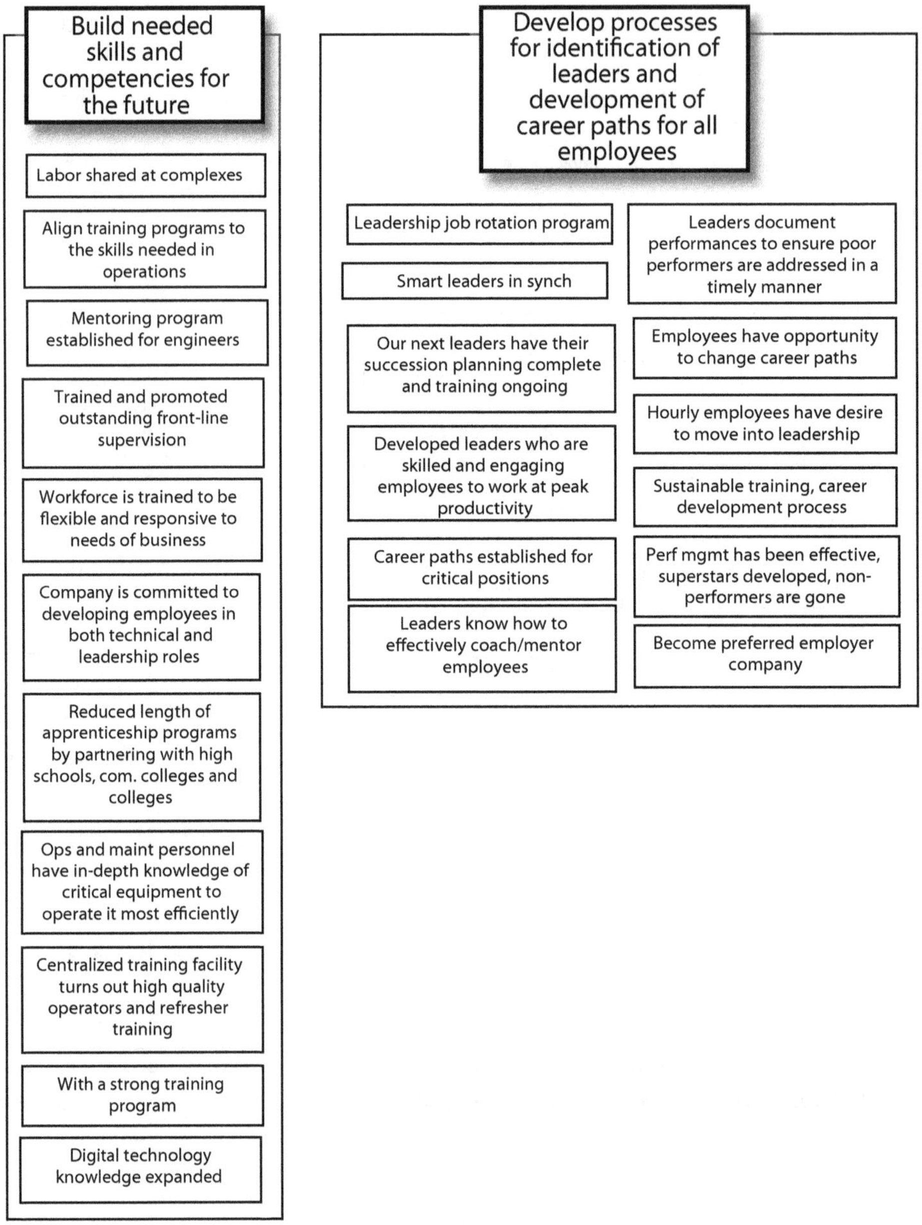

Figure A.1 (Continued) Vision Affinity Diagram case study.

Interrelationship Digraph

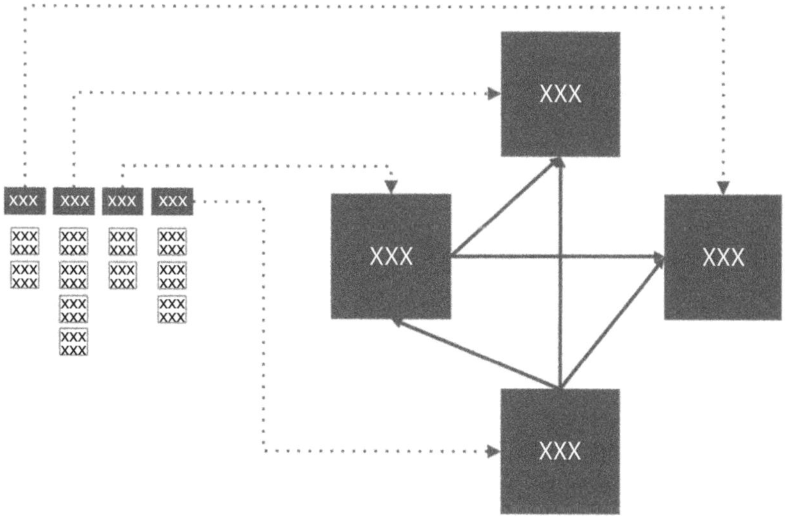

Tool Purpose

The Interrelationship Digraph is used to show the relationships among different parts of a system. In organizational settings, the Interrelationship Digraph is frequently populated by the header cards from the Affinity Diagram.

Tool Mechanics

The header cards from the Vision Affinity Diagram are placed in a circle, and for each relationship between two items, the leadership team asks, "Is there a relationship between these two items?" and "Which drives or influences the other most?" Hoshin planners draw arrows between each set of two items to determine which item is driving the others. An item with many outgoing arrows is a "driver," and an item with many incoming arrows is an "outcome." Drivers are candidates for being chosen as Hoshin objectives. Outcomes generally reflect an area to place a metric that will show the results of work on the drivers.

Uses in the Hoshin Planning Cycle

The Interrelationship Digraph is used as a first round of analysis in choosing the annual Hoshin objective. Leaders consider the items with the most outgoing arrows as potential Hoshin areas. In addition, they might discuss the item with the most incoming arrows when they define high-level metrics or track the organization's performance. Those items are most impacted or driven by the work in other areas.

Figure A.2 continues the power generation company case example. Notice how the items in the Interrelationship Digraph are taken directly from the case study Affinity Diagram header cards in Figure A.1.

Considerations

- This tool is demanding, and it's best not to stop a leadership team partway through its construction. Schedule enough time to finish.
- When fatigue sets in, particular care should be taken to differentiate between real consensus and people simply giving up on the dialogue.
- The item with the most outgoing arrows isn't always chosen as the Hoshin. Leaders apply judgment to thinking through all the work done so far and making a thoughtful choice.

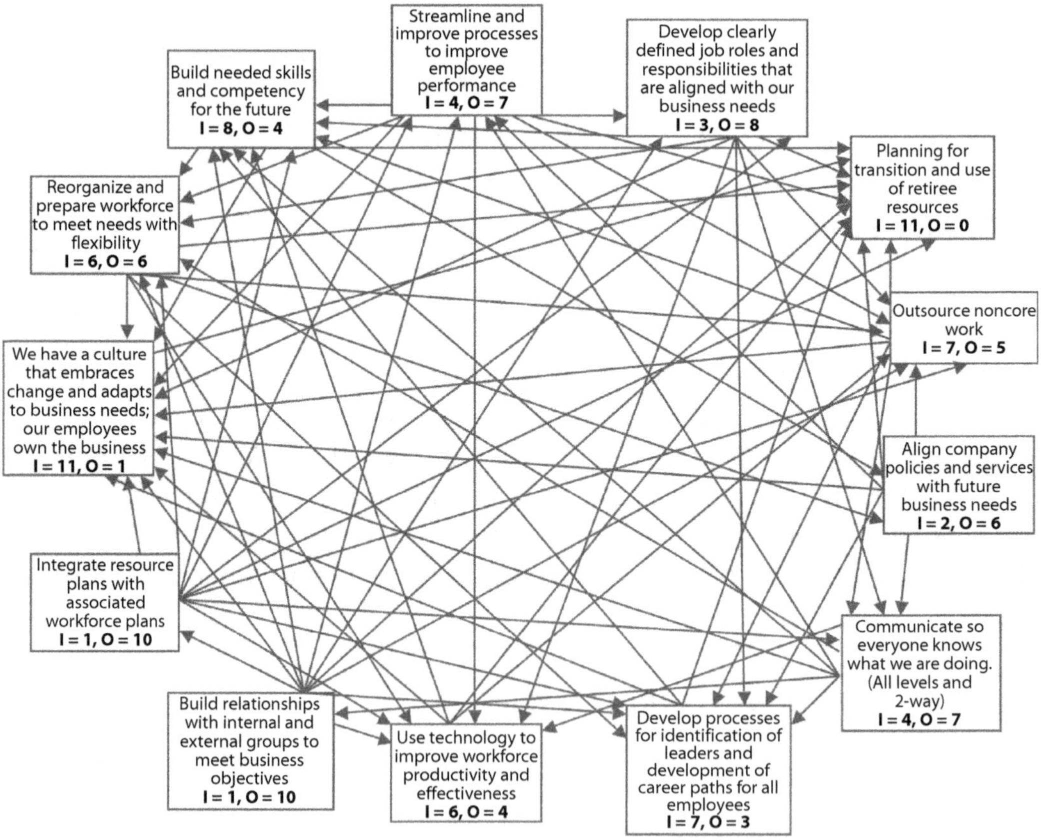

Figure A.2 Interrelationship Digraph, case study.

Radar Chart

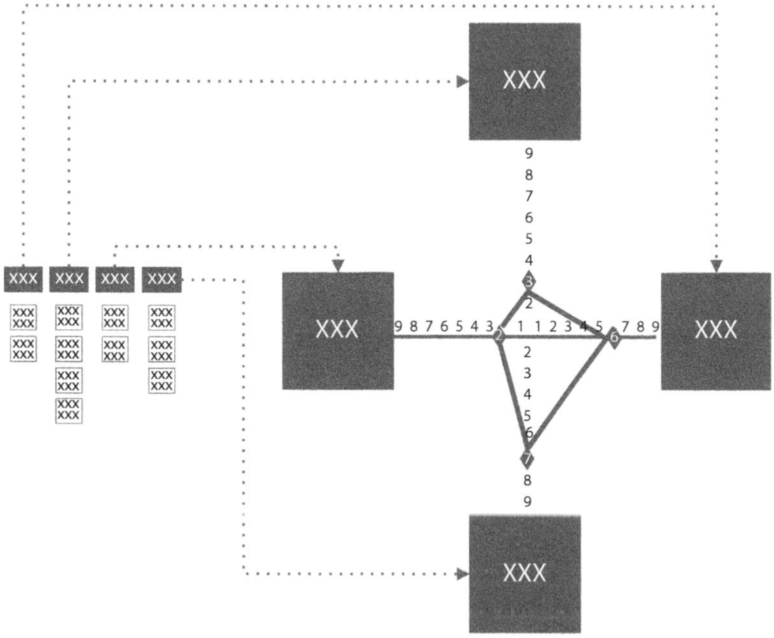

Tool Purpose

The Radar Chart is used to evaluate the current performance of the elements of a system. A scale is used, often from 1 to 10, to show relative performance among the elements. Used in a group setting, the "score" for each element must be arrived at through consensus. The discussion that takes place to arrive at a number is as valuable as the score itself.

Tool Mechanics

Like the Interrelationship Digraph, the Radar Chart is populated with the header cards from the Vision Affinity. On a scale, usually 1 to 10, each header card area is evaluated for how well the organization is currently accomplishing it. Leadership teams may either discuss until they arrive at a consensus score for each item or rate the items individually and combine the scores.

Uses in the Hoshin Planning Cycle

The Radar Chart is used to help a leadership team discuss and evaluate its current performance on the main components of the Vision Affinity. It is used as an adjunct to the Interrelationship Digraph. Together the two tools help leaders hone in on a leveraged choice for that year's Hoshin. Areas that surface as drivers in the Interrelationship Digraph and that are not currently done very well by the organization may be good candidates for a Hoshin objective.

Figure A.3 shows the Radar Chart from our case study example. Each of the areas represented on the header cards used in the Interrelationship Digraph in Figure A.2 is evaluated for how well the organization is currently performing it. The further out on the line (the higher the number), the better the leadership team believes the organization currently is at doing it.

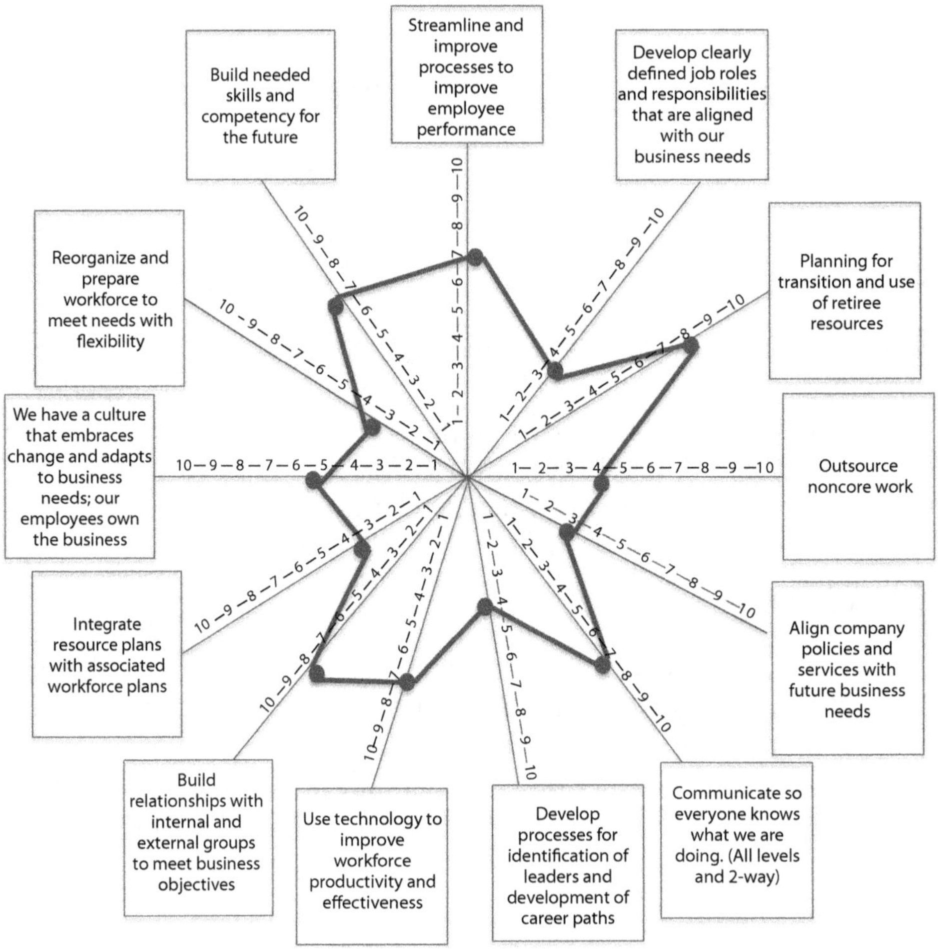

Figure A.3 Radar Chart, case study.

Considerations

- If the scale is 1 to 10, and a team starts assigning everything a 5, that won't contribute as well to the selection of the Hoshin objective. It can help to consider the performance in each area relative to the performance in other areas. For example, teams might ask, "Are we better at this area than this one? How much better?"
- If a consensus score is elusive, leaders can try first having each person rate the items individually, then sharing, one at a time, with the group how he or she arrived at that score.

Annual Plan Table

Hoshin objective	Strategy	Owner	Metric	Goal
	1.1			
Metrics	2.0			
	3.0			
Goals	4.0			
	5.0			

Tool Purpose

The Annual Plan Table is used to organize on one page the main components of the Hoshin plan. Leadership teams can see, in one place, the Hoshin objective, the strategies to achieve it, the metrics that will be used to assess progress toward the objective, and the people who are accountable for each strategy. This table is often used as a simple summary to communicate the Hoshin plan to the next levels of leadership in the organization during Catchball, as well as to Boards of Directors and other stakeholders.

Uses in the Hoshin Planning Cycle

The Annual Plan Table is used to express the highest level of the Hoshin plan, as well as the second and maybe its third and fourth levels. Teams switch to the Implementation Table when the metrics being assigned are mostly dates and when the tactics listed may be assignable to one person. As the plan is broken down, the words "strategies" and "tactics" may be seen instead of "objective" and "strategies."

In the example of the Annual Plan Table in Figure A.4, three related items from the Interrelationship Digraph and Radar Chart have been combined to form a Hoshin objective: "Reorganize and prepare workforce to meet needs with flexibility," "Develop clearly defined job roles

Hoshin Year: 2011/2012

Situation: Bleak economy, heightened attention to carbon emissions, aging workforce.

Hoshin Objective: Build Our Workforce for the Future

Hoshin	Strategies	Owner	Measures	Goals
Create a system to continuously define and and produce competencies and skills that meet business needs	1.0 Model likely future situation: mix of power generation assets, relationships with Public Utility Commission, legislators, etc.	L.G.	• 2020/ 2030 generation assets modeled	• By 8/11
	2.0 Create a process for continuously reviewing and producing current and future competencies and staffing and build staffing design for 2020.	M.A.	• 2020 staff design • Gap analysis — current & 2020 • % core & non-core work defined • % job descriptions done for 2020	• 11/12 • 12/11 • 100% • 100%
Measure (s)	3.0 Develop Service Level Agreements (SLAs) with internal/external service providers.	A.S.	• % SLAs identified and drafted • Review process and remedies in plac	• 12/11 • 2/12
1. % defined staff needs met 2. % workforce > "3" on engagement survey (scale 1-5) 3. Competency test scores 4. % non-core work by external labor	4.0 Develop multi-year, phased labor strategy for 2020 workforce and align with contract negotiation schedule.	B.A.	• % labor contract negotiations including future workforce discussion	• 100%
	5.0 Develop a skills-based training plan for our power generation 2020 workforce.	T.R.	• % of jobs defined by team 2.0 that have a training plan established. • Evaluation of different means of skill-training.	• 100% • 6/12
Goal	6.0 Create a *two-way* communication process to develop and implement the core competencies for 2020.	A.L.	• # of Face-to-Face, Focus Groups • % survey scores > 3 on scale 1 to 5	• 1 per quarter • 100%
1. 80–90% by 12/12 2. 100% 3. Top 10th industry percentile nationally 4. 100%	7.0 Identify emerging technologies and technology strategy with the 2020 (and beyond) workforce.	P.M.	• Gap analysis • Benchmark study	• Both by 12/12

Figure A.4 Annual Plan Table, case study.

and responsibilities that are aligned with our business needs," and "Align company policies with future business needs." The Hoshin is called *Build Our Workforce for the Future.* Those three items were among the drivers on the Interrelationship Digraph and were also low scorers on the Radar Chart, and they were all related to developing the human resources of the organization.

A plan to complete the Hoshin *Build Our Workforce for the Future* was created using an Affinity Diagram (not shown). The headers from that Affinity are reflected in the Annual Plan Table in Figure A.4 as 1.0, 2.0, 3.0, and so forth.

Considerations

- There is intention behind the one-page design of the Annual Plan Table. Keeping to one page forces an economy of words and numbers that leads the planner to choose descriptors thoughtfully and can allow for efficient communication of the plan at meetings.

Implementation Table

Hoshin Objective	Tactics	Owner	Q1	Q2	Q3	Q4
	1.0					
Metrics	2.0					
	3.0					
Goals	4.0					
	5.0					

Tool Purpose

The Implementation Table is used to organize on one page one subsection of the Hoshin plan that has been broken down to the level of assignable tasks. Because it includes bars representing amounts of time to complete a task, it can be good input for use in building an overall project schedule. The bars representing time resemble another tool called a Gantt Chart, which is sometimes referred to in relation to the Implementation Table.

Uses in the Hoshin Planning Cycle

The Implementation Table is used to express more detailed levels of the Hoshin plan. Teams may switch to the Implementation Table when they need a visual representation of the schedule of their work at a detailed level or when the tactics identified are assignable as discrete tasks. The time for moving from the Annual Plan Table to the Implementation Table can also present itself when plan owners notice that the metrics being assigned are mostly dates.

Software packages often combine the Gantt feature with the Annual Plan Table and the Review Table in one overall visual tool. Even if software will be used, there is some value to considering the three core Hoshin Planning tables as distinct from each other when the plan is being formed. The thinking that goes into each type is different. For example, more strategic and team discussion takes place around the Annual Plan Table, more discussion takes place around the Implementation Table level.

Date: 2011	Strategy Owner: J.F. Dufresne	Strategy: 2.0 Create process for reviewing and producing competencies and staffing, and build staffing design for 2020. Sub-Strategy: 2.5 Create 2020 staffing design.	Hoshin Objective: Build Our Workforce for the Future

Strategy	Tactics/Implementation Items	Owner	Q3 - 11	Q4 - 11	Q1 - 12	Q2 - 12
2.5 Create the 2020 staffing design	**2.5.1** Identify future core competencies needed		6-1-12			
Measures	**2.5.2** Create onboarding and training policies to support new staffing design		6-1-11			
• 2020 staff design • Gap analysis — current & 2020 • Core & non-core work defined • % job descriptions. done for 2020	**2.5.3** Assess current staff and competencies		12-11			
	2.5.4 Policy created for periodic asset mix impact review			3-12		
Goals • By 11/12 • By 12/11 • 100% • 100%	**2.5.5** Provide a draft of the staffing design for 2020					*11-12* ->

Figure A.5 Annual Plan Table, sub-strategy implementations 2.5, case study.

Figure A.5 continues our power generation company case study example with a breakdown of the 2.0 area seen in the Annual Plan Table in Figure A.4. The 2.0 strategy is broken down again into sub-strategies 2.1, 2.2, 2.3, 2.4, 2.5. For each of those sub-strategies, there is an Implementation Table like this one in Figure A.5. Only the table for 2.5 is shown here as an example.

Considerations

- More detailed plans provide a better vehicle for leaders to assess what the resource impact on the organization of the Hoshin will be. Stopping too soon can leave a vagueness at the implementation level of the plan that can cause things to not get done.

Review Table

<table>
<tr><td>Review Date:</td><td>Strategy Owner:</td><td colspan="2">Strategy:
Sub-Strategy:</td><td colspan="2">Hoshin Objective:</td></tr>
<tr><th>Strategy or Tactic (Plan)</th><th>Goal (Plan)</th><th>Actual Performance (Do)</th><th>Made/Missed ●/○ (Check)</th><th>Root Cause of Deviation (Check)</th><th>Corrective Action (Act)</th></tr>
<tr><td>1.0</td><td></td><td></td><td></td><td></td><td></td></tr>
<tr><td>2.0</td><td></td><td></td><td></td><td></td><td></td></tr>
<tr><td>3.0</td><td></td><td></td><td></td><td></td><td></td></tr>
<tr><td>4.0</td><td></td><td></td><td></td><td></td><td></td></tr>
<tr><td>5.0</td><td></td><td></td><td></td><td></td><td></td></tr>
</table>

Tool Purpose

The Review Table is used to report on the progress of the Hoshin plan between and during review meetings. If it is not automated with software, people reporting out create a slide of the Review Table that highlights what metrics were made and which ones were missed. Review Table shows the typical template used.

Uses in the Hoshin Planning Cycle

The Review Table is structured to make the Plan–Do–Check–Act (PDCA) cycle real. Notice that there is at least one column reflecting each of the four phases of the PDCA cycle. The consistent use of the Review Table helps ensure that the cycle is completed and that discussions around the plan's progress are explicit in terms of root causes of plan breakdowns.

Figure A.6 closes out our case study example with a Review Table on sub-strategy 2.5. Note how for each work activity shown in the Implementation Table in Figure A.5, its status is shown as either "Made" or "Missed," and all "Missed" items note a cause and a planned or completed corrective action.

Review Date: 12/8/11	**Strategy Owner:** J.F. Dufresne	**Strategy:** 2.0 Create a process for reviewing and producing competencies and staffing, and build staffing design for 2020. **Sub-Strategy:** 2.5 Create 2020 Staffing Design			**Hoshin Objective:** Build Our Workforce for the Future	
Strategy or Tactic (*Plan*)		**Goal (*Plan*)**	**Actual Performance (*Do*)**	**Made/Missed ●/○ (*Check*)**	**Root Cause of Deviation (*Check*)**	**Corrective Action (*Act*)**
2.5.1 Identify future core competencies		50% by 12/11 100% by 6/12	≈ 10%	○	Access to benchmarking data limited. No deliverables from strategies 1.0 and 7.0	Revised benchmark organization list. 2.5 member attending 1.0/9.0 team meetings
2.5.2 Create onboarding process to support new staffing design		Policy formalized 5/12	Policy completed and in review	●		
2.5.3 Assess current staff and competencies		All 11 plants studied by 12/11	6 of 11 plants studied	○	Underestimated time needed at each plant	Revised plans for remaining 5 plants to include more pre-work for plant supers
2.5.4 Create policy for periodic asset mix impact review		Policy formalized 3/12	Policy completed and in review	●		
2.5.5 Provide a draft of the staffing design for 2020		Draft design by 11/12	On target with activities	●		

Figure A.6 Review Table on sub-strategy 2.5, case study.

Considerations

- Dedicating meeting time to the items in the plan that are "missed" can make the best use of time. The missed items may be where more learning is available and more plan adjustment is needed.
- Create clear, high-quality metrics in the Annual Plan Tables. This allows for more objective assessment of how well the implementation of the Hoshin is going later during review. Vague or missing metrics can make an objective assessment of the plan's progress difficult or impossible.

Catchball

Tool Purpose

Catchball is used to solicit and organize feedback on a plan. In the classic images of Catchball, the visual metaphor of a baseball is used to show how the plan is tossed to the next level of management, and then tossed back up with feedback attached to it. In reality, what is tossed back and forth is the Annual Plan Table that has been populated by the leadership team with the strategies, metrics, and names of strategy leaders for accomplishing the Hoshin objective. See Figure A.7 for a sample Catchball.

Uses in the Hoshin Planning Cycle

Once the highest level plan is developed, Catchball is used to check whether metrics and strategies are realistic, to finalize the team of strategy owners, and to make a practical assessment of the impact of the Hoshin plan on the organization's resources. On Catchball, the main strategies to achieve the Hoshin are designated as 1.0, 2.0, 3.0, and so forth and are shared with division or department level managers, usually via meeting discussions. Formal input on the plan is solicited.

Those who have been assigned a Hoshin strategy then employ Catchball to solicit feedback on the Annual Plan Table their team has created for their strategy. On Catchball, for example, the 3.0 strategy team will create a plan at that level, and then review it with the people completing items 3.1, 3.2, 3.3, and so forth as well as other stakeholders related to strategy 3.0.

The actual process of Catchball includes meetings where the plan is presented, resource needs are considered, how the plan fits with other work in the organization is discussed, and reality checks are applied to the metrics in the plan. The use of software to share and document suggested changes to the plan probably makes the mechanics of Catchball a little easier at this stage, but simply sharing plans electronically and soliciting feedback should not be used as a substitute for real dialogue.

Catchball may surface skill gaps in the organization's capability to accomplish the strategy, the need for different or more people on the team from across the organization, and ideas for how to accomplish the strategy differently.

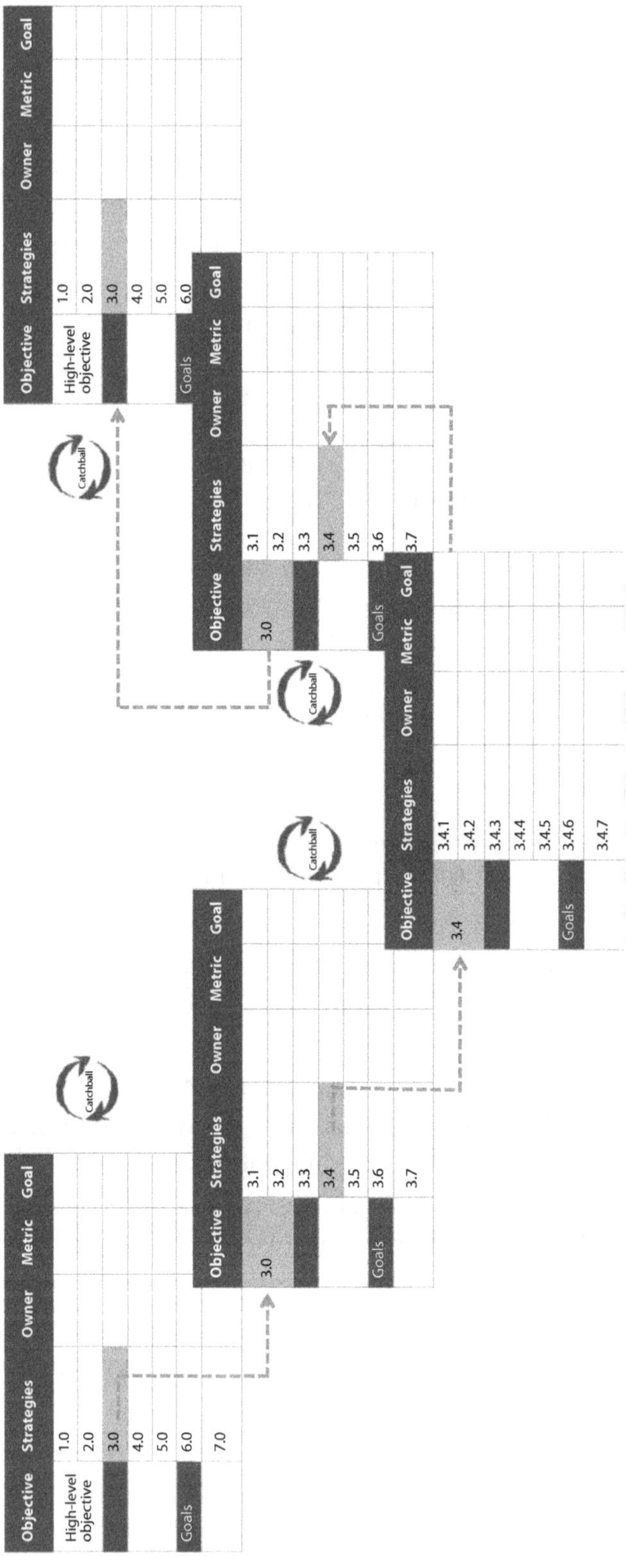

Figure A.7 Catchball.

Considerations

- The highest level Annual Plan Table should be presented by a member of the leadership team and include an explanation of the strategic context and work that led to the selection of the Hoshin. Catchball meetings are early practice opportunities for leaders in articulating the logic of the plan.
- Feedback on the plan should be documented formally, even if the meetings are kept casual. This is an early opportunity to move people toward the rigorous level of documentation most typical of well-practiced Hoshin Planning.
- Conduct Catchball at all levels of the plan, if it is practicable to do so. The idea of testing the assumptions in a plan before implementing it makes good sense at any level.

Pareto, Fishbone, and Run Chart

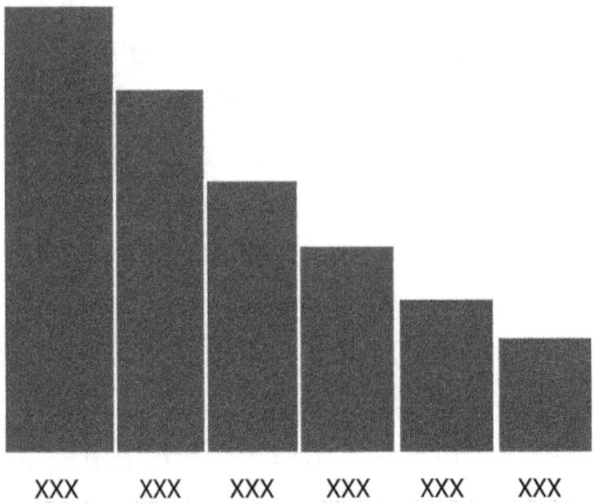

(a)

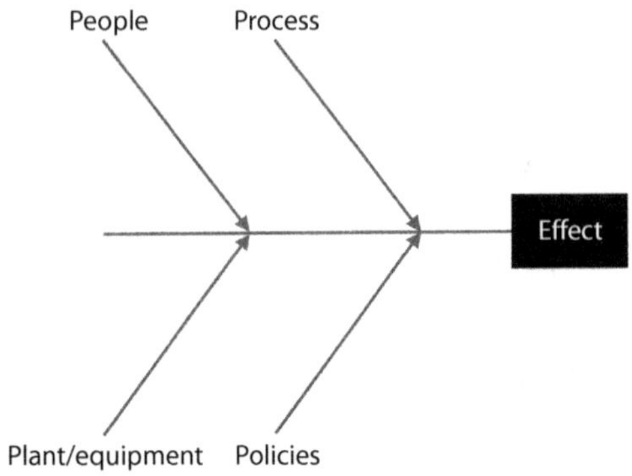

(b)

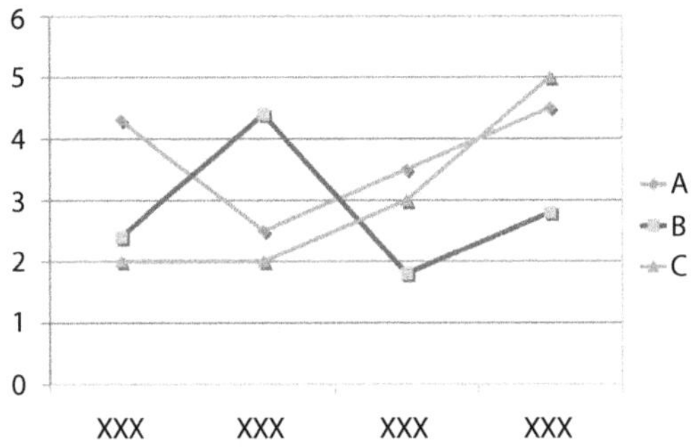

(c)

Tools Purpose

Pareto Charts (a), Fishbone Diagrams (b), and Run Charts (c) are on a menu of data processing tools, which together are used for Quality Control (QC) or Statistical Process Control (SPC).

Pareto Charts are used to identify the most frequently occurring problem or cause in a system, to focus resources on problems that offer the greatest potential for improvement. They look like bar charts, organized in descending order. The x-axis represents the categories of causes or problems, and the y-axis represents the frequency of its occurrence.

Fishbone Diagrams are used to show all the possible causes of a problem. Those studying a problem place the "effect" on the right side of the chart and brainstorm its causes to the left. The classic categories for the causes are the "Four M's" and the "Four P's."

Four M's and Four P's

Four M's

- Manpower
- Machinery
- Methods
- Materials

Four P's

- People
- Processes
- Plant and Equipment
- Policies

Run Charts are used to track data over time, where the x-axis represents time. Reading Run Charts, Hoshin planners look for trends, cycles, and shifts in important performance data to indicate changes, for better or for worse, in the processes that deliver value to customers.

Use in the Hoshin Planning Cycle

Pareto Charts, Fishbone Diagrams, and Run Charts are featured primarily in the review phase of a Hoshin Planning cycle. At the start of Hoshin, executive teams may use these and other QC tools to display a current assessment of the organization's performance on key measures and the reasons for any shortcomings.

During regular Hoshin review meetings, these tools are used to explain any "Missed" metrics. For all those open circles shown in the Review Table given earlier, strategy owners must conduct a root cause analysis, showing the reasons the goal was missed (Fishbone) and data to support the conclusion (Pareto and other data tools).

In addition, Figure 4.7 in Article 4, the measurement article in this book, refers to the Flag Chart. The overall structure of the Flag Chart resembles a Fishbone Diagram in how it breaks a problem down into more and more detailed causes. The "flags" themselves often represent the data from a Pareto Chart, showing where the biggest concentration of the problem sits.

Bibliography

Akao, Yoji. *Hoshin Kanri: Policy Deployment for Successful TQM.* Cambridge, MA: Productivity Press, 1991.

Baden-Fuller, Charles, and Haefliger, Stefan. "Business Models and Technological Innovation." *Long Range Planning*, 46: 419–426, 2013.

Bartz, Nick, Cassels, Tom, Egan, Yulan, and Kerns, Christopher. *Best-in-Class Integration of Newly Acquired Entities.* The Advisory Board Company, 2013.

Boisvert, Lisa, Brassard, Michael, Hotopp, Dona and Ritter, Diane. *Coach's Guide to The Memory Jogger™ II.* Salem, NH: GOAL/QPC, 2010.

Boisvert, Lisa. "Strategic Planning Using Hoshin Kanri." *Executive White Paper Series.* Salem, NH: GOAL/QPC, 2011.

Brassard, Michael, and Ritter, Diane. *The Memory Jogger™ II.* Salem, NH: GOAL/QPC, 2010.

Bridges, William, and Bridges, Susan. *Managing Transitions: Making the Most of Change.* New York: De Capo Lifelong Books, 2009.

Cain, Susan. *Quiet: The Power of Introverts in a World That Can't Stop Talking.* New York: Broadway Books, 2013.

Chase, Robin. *Peers, Inc: How People and Platforms are Inventing the Collaborative Economy and Reinventing Capitalism.* New York: PublicAffairs, 2015.

Cowley, Mike, and Domb, Ellen. *Beyond Strategic Vision: Effective Corporate Action with Hoshin Planning.* New York: Routledge, 2011.

Deming, W. Edwards. *Out of the Crisis.* Cambridge, MA: MIT Press, 1982.

DePree, Max. *Leadership Is an Art.* New York: Dell Publishing, 1989.

Gazzaniga, Michael. *Human: The Science Behind What Makes Us Unique.* New York: Ecco HarperCollins, 2008.

Geisst, Charles R. *Deals of the Century.* Hoboken, NJ: John Wiley & Sons, 2004.

Gottschall, Jonathan. *The Storytelling Animal: How Stories Make Us Human.* New York; Houghton Mifflin Harcourt, 2012.

Ishikawa, Kauro. *Guide to Quality Control.* White Plains, New York: Quality Resources.

Jackson, Thomas. *Hoshin Planning for the Lean Enterprise.* New York: Taylor & Francis, 2006.

Knott, Anne Marie. "The Trillion-Dollar R&D Fix." *Harvard Business Review*, May 2012.

Magretta, Joan. *Understanding Michael Porter: The Essential Guide to Competition and Strategy*. Cambridge, MA: Harvard Business Review Press, December 6, 2011.

Melum, Mara Minerva. *Breakthrough Leadership: Using Hoshin Planning*. Boston: American Hospital Association and GOAL/QPC, 1998.

Porter, Michael E. "The Five Competitive Forces That Shape Strategy." *Harvard Business Review*, January: 78–93, January 2008.

Porter, Michael E., and Lee, Thomas H. "Why Strategy Matters Now." *New England Journal of Medicine*, April 2013.

Porter, Michael E., and Lee, Thomas H. "The Strategy That Will Fix Healthcare: Providers Must Lead the Way in Making Value the Overaching Goal." *Harvard Business Review*, October 2013.

Project Management Institute. *A Guide to the Project Management Body of Knowledge*. 2013. Newtown Square, PA: Project Management Institute.

Rosenzweig, Phil. *The Halo Effect*. New York: Free Press, 2007.

Robert, Michel. *The New Strategic Thinking*. New York: McGraw-Hill, 2005.

Schein, Edgar H., and Bennis, Warren G. *Personal and Organizational Change Through Group Methods*. New York: John Wiley & Sons, 1965.

Schwarz, Charles. *The Skilled Facilitator*. San Francisco: Jossey-Bass, Publishers, 1994.

Steinbrenner, Peter, and Boisvert, Lisa. "Using Hoshin Planning to Redesign a Workforce." *Electric Energy*, Fall, 18–24, 2012.

Weick, Karl E., and Sutcliffe, Kathleen M. *Managing the Unexpected: Sustained Performance in a Complex World*. San Francisco: Jossey-Bass.

Index

A

O

P

Q

R

S

T

www.ingramcontent.com/pod-product-compliance
Lightning Source LLC
LaVergne TN
LVHW080652011026
849160LV00014B/1398
9781482299786